W9-CSP-149

Berkeley

PROCEEDINGS OF THE SYMPOSIUM ON

HIGH POWER, AMBIENT TEMPERATURE LITHIUM BATTERIES

Edited by

W. D. K. Clark
Wilson Greatbatch Ltd.
Clarence, New York

Gerald Halpert
Jet Propulsion Laboratory
California Institute of Technology
Pasadena, California

BATTERY DIVISION

Proceedings Volume 92-15

THE ELECTROCHEMICAL SOCIETY, INC., 10 South Main St., Pennington, NJ 08534-2896

ɞ 4515791

CHEMISTRY

Copyright 1992

by

The Electrochemical Society, Incorporated

*Papers contained herein may not be
reprinted and may not be digested by publications other than those of The Electrochemical
Society in excess of 1/6 of the material presented.*

Library of Congress Catalog Number: 92-81315

ISBN 1-56677-014-9

Printed in the United States of America

*TK2921
S88
1991
CHEM*

PREFACE

While aqueous based battery systems have traditionally been used for applications needing high power densities, lithium battery systems have evolved to the point where they overlap with these and high temperature battery systems in terms of their ability to deliver power. At the Fall Meeting of the Electrochemical Society in Phoenix, Arizona, October 13-17, 1991, papers were presented in a symposium which focussed on all aspects of lithium systems as they relate to the performance, shelf life and safety of high power lithium systems. The first paper of the symposium presented a pointed comparison between the liquid cathode lithium systems and the silver-zinc battery, the traditional high power ambient temperature power source. The authors show that these lithium systems have demonstrated power levels of the same order as the silver-zinc and even thermal battery systems but offer significant advantages in energy density and and wet-stand time.

While the primary systems as noted above are close to aqueous systems in terms of their power levels, the secondary systems have to make significant improvements in this aspect of their performance to approach the aqueous systems. However, the papers on the secondary systems presented improvements in many elements such as newer structures for the lithium-insertion anodes, work on the organic electrolytes and the electrolyte/cathode interface, and the continued study of the polymer electrolyte systems with their potential for thin film fabrication. Improvement in power levels for the rechargeable systems was also examined through consideration of design elements such as bipolar configurations. While probably never being able to surpass the aqueous systems in terms of power density, lithium battery systems have made significant strides over the past decade and offer more viable solutions in many applications where power is not the overriding factor.

We wish to thank the authors for the time spent in preparation of the papers and for the excellent presentations during the symposium in Phoenix. We also wish to express our thanks for the patience and guidance of the publications staff at the Electrochemical Society Headquarters.

W. D. K. Clark
Wilson Greatbatch Ltd.
Clarence, New York

G. Halpert
Jet Propulsion Laboratory
California Institute of Technology
Pasadena, California

iii

CONTENTS

v

vi

FACTS ABOUT THE ELECTROCHEMICAL SOCIETY, INC.

The Electrochemical Society, Inc., is a nonprofit, scientific, educational, international organization founded for the advancement of the theory and practice of electrochemistry, electrothermics, electronics, and allied subjects. The Society was founded in Philadelphia in 1902 and incorporated in 1930. There are currently over 5000 scientists and engineers from more than 40 countries who hold individual membership; the Society is also supported by more than 100 corporations through Patron and Sustaining Memberships.

The technical activities of the Society are carried on by Divisions and Groups. Local Sections of the Society have been organized in a number of cities and regions.

Major international meetings of the Society are held in the Spring and Fall of each year. At these meetings, the Divisions and Groups hold general sessions and sponsor symposia on specialized subjects.

The Society has an active publications program which includes the following.

JOURNAL OF THE ELECTROCHEMICAL SOCIETY - The JOURNAL is a monthly publication containing technical papers covering basic research and technology of interest in the areas of concern to the Society. Papers submitted for publication are subjected to careful evaluation and review by authorities in the field before acceptance, and high standards are maintained for the technical content of the JOURNAL.

EXTENDED ABSTRACTS - Extended abstracts of the technical papers presented at the Spring and Fall Meetings of the Society are published in serialized softbound volumes.

PROCEEDINGS VOLUMES - Papers presented in symposia at Society and Topical Meetings are published from time to time as serialized softbound Proceedings Volumes. These provide up-to-date views of specialized topics and frequently offer comprehensive treatment of rapidly developing areas.

MONOGRAPH VOLUMES - The Society has, for a number of years, sponsored the publication of hardbound Monograph Volumes, which provide authoritative accounts of specific topics in electrochemistry, solid state science and related disciplines.

STATE OF THE ART IN HIGH POWER LITHIUM BATTERIES

P. Chenebault and J.P. Planchat

SAFT ABD

Rue Georges Leclanché

BP 1039

86060 POITIERS Cedex

FRANCE

Abstract:

This paper presents an overview of the liquid cathode systems used in high rate batteries. After a description of the advantages of these systems compared to other lithium systems, the electrochemical limitations under high rate discharge conditions are discussed. These considerations are then illustrated by practical examples representing the state of the art in the field of high power lithium batteries. Other potentially substitute systems are then presented.

INTRODUCTION

In the field of high power batteries, the liquid cathode systems receive considerable interest because of their high energy density and high rate capabilities. In reserve type configurations, these batteries provide alternatives to more "classical" systems (ie AgO/Zn, AgO/Al). Their possible use for missile or underwater applications has led to the development of high power batteries.

After a brief description of the advantages of the liquid cathode systems among other lithium systems and an analysis of the electrochemical limitations, this paper presents the state of the art in the field of high power lithium batteries. A presentation of related systems under investigation is made.

SPECIFICITY OF HIGH RATE LITHIUM BATTERIES

Lithium batteries can be classified with respect to their discharge temperature and configuration. The highest rate capabilities with the lithium anodes, up to several amperes/cm^2, can be obtained at high temperature (600°C) [1]. High temperature discharge, however, generally limits discharge time, and it results in a continuous

1

voltage decrease due to reduced electrolyte conductivity with lower temperature.

Ambient temperature lithium batteries are generally categorized based on the type of electrolyte and cathode.

Solid cathode systems do not offer high rate capabilities mainly because of the low conductivity of the electrolyte, but also because of the poor kinetics of the cathodic reduction occuring in the solid phase. Li/V_2O_5 is the only known system used in a small reserve type configuration for mine fusing, but the discharge current density is rather low (a few mA/cm^2) [2].

Active liquid cathode batteries ($Li/SOCl_2$ or Li/SO_2Cl_2) can be discharged at 5 mA/cm^2 to 10 mA/cm^2, (discharge under constant 60 mA/cm^2 have been reported on $li/SO_2Cl_2 + Cl_2$ system [3]). The main limitations to their use under high rate conditions comes from the passivating layer of lithium chloride and from their internal configurations.

The spiral configuration of these cells limits heat exchange from the center to the periphery of the cell, leading to overheating of the cells when discharged under high drain [6]. Moreover, internal fuses are commonly included in the cells to prevent discharge under too high rate.

To overcome the voltage delay and selfdischarge problems associated with the growth of the passsivating layer on the lithium, a "special electrolyte" with a rather low $LiAlCl_4$ concentration (between 0.5 and 1.5 molar) and proprietary additives is used [4],[5].

In reserve type batteries, liquid cathode system can be used under high rate conditions (up to 500 mA/cm^2) using either thionyl chloride or sulfuryl chloride with concentrated electrolyte (acidic or neutral) and with different salts and additives [7][8][9]. In this configuration, no self discharge occurs during storage and the initial voltage is not affected by the voltage delay phenomenon. Moreover, discharge conditions such as low temperature and wet stand time can be supported by these cells without degradation of their performance.

Compared to others conventional electrochemical systems used in similar high power applications, the liquid cathode reserve type batteries offer the following advantages:

-Extended discharge temperature versus AgO/Zn (no preheating needed).

-Possibility of longer discharge duration versus AgO/Zn, AgO/Al, thermal batteries.

-Possibility of longer wet stand time versus AgO/Zn, thermal batteries.

-Long shelf life

-Flat discharge plateau

-Higher energy density versus classical systems

The main applications are power supplies for missile, and underwater vehicles, which combine optimized electrochemistry with a specific design for a high reliability.

Because of the generally short discharge duration of these batteries, improvement of the energy density is possible by the use of light weight materials which are stable in the sulfur oxychloride solvents for several days. New polymers and composites present opportunities to save weight.

ELECTROCHEMISTRY

A large number of description of $Li/SOCl_2$ or Li/SO_2Cl_2 liquid cathode systems, have been published [4],[10],[11]. A complete description of their operation exceeds the scope of this paper. We focus here on specific problems associated with high rate discharge conditions.

The discharge voltage, V, of a cell corresponds to the following equation:

$$V = E° - \eta_A - \eta_C - RI$$

where E° is the Emf, η_A and η_C the respecte anodic and cathodic polarization and RI the ohmic losses. A high discharge voltage requires a maximum E° and minimized polarizations and ohmic losses.

Anodic polarization

The anodic polarization losses are generally negligeable compared to ohmic losses and cathodic polarization. The most interesting papers published on that subject are those of Hagan [12] and James [13].

During discharge of different lithium samples in $SOCl_2$ based electrolyte, Hagan obtained significant polarization variation above 100 mA/cm^2 (up to 100 mV from one sample to an other). They attributed this to differences in impurities levels from one sample to another which influences the morphology and electronic conductivity of the passivating layer of lithium chloride.

The rapid built up of the LiCl surface layer promotes a phenomenon observed by James under high current density (from 100 to 1800 mA/cm^2) which is caracterized by a rapid voltage drop. The time to reach this drop is reduced with increasing current density. This phenomenon results from accumulation of anodic

3

dissolution products in the pores of the surface layer which block the surface. Nevertheless, it is reversible and the discharge can be continued after a small rest period. Similar behavior has been reported at lower current density for the calcium anode, material which tends to passivate more rapidly than lithium in these electrolytes [14]. The study by James of the influence of $AlCl_3$ addition to limit precipitation showed that it is not obvious because of competition between rates of LiCl and lithium ion formation as well as neutralization of the LiCl. This precipitation can lead to anodic limitation with very high overpotential and associated cell overheating.

The quality of the "natural" oxide on the lithium surface can also influence its electrochemical behaviors, as does the LiCl layer. No specific studies have been made on this topic, certainly because of the difficulty of to controlling that layer.

The polarization of the electrode may also result from other factors like the drying out of the lithium/separator interface which can be overcome by the use of very porous wetable separators. The texture of some woven separators can also affect the polarization by creating a mask effect leading to higher current density concentrations in the open areas of the separators.

Electrolyte

The electrolyte formulation results from combination of:
cathodic material ($SOCl_2$, SO_2Cl_2)
$\qquad$ -salts ($LiAlCl_4$, $LiGaCl_4$)
$\qquad$ -additives ($AlCl_3$, SO_2)
which is an optimization based on compromises between characteristics like viscosity, conductivity, density, surface tension, vapour pressure, discharge performance.

A very large number of electrolyte formulations have been studied and a usefull database has been established [15].

Compared to aqueous solvents used for example in AgO/Zn batteries, the conductivity is quite low (20 mS/cm compared to 600 mS/cm for a 30% KOH aqueous solution). This leads to significant ohmic losses which are overcome by the use of very thin, porous and wetable fiber glass separators. On the other hand, in the reserve batteries, this low conductivity enables the possibility of long wet stand time by minimizing leakage currents and their side effects.

Nevertheless, these electrolyte still have acceptable conductivity at temperatures well below 0°C providing for the possibility of discharge without preheating.

4

The structure of these electrolytes has been studied from a macroscopic point of view by Raman spectroscopy [16], showing the formations of solvates between lithium ion and SO_2 and $SOCl_2$. These results give useful information on the electrolyte evolution during the discharge.

Microscopic studies have been made to determine the type of bond in molecules of adducts or solvates [17]. These results provide data for the study of the cathodic charge transfer process .

Derivative electrolytes realized with solvates like $LiAlCl_4$, 3 SO_2 give higher conductivity at ambient temperature (100 mS/cm = 5 times that of standard thionyl chloride solutions). Used for lithium/SO_2 rechargeable systems [18], these solutions also appear to be particularly suitable to study halides and oxyhalides reduction [19]. Similar solutions have been used recently for high rate battery discharged in a wide temperature range (-30°C to + 70°C) [20], and then can be very usefull for a large number of applications requiring pulses.

Cathodic polarization

The cathodic polarization represents the main part of the overall polarization and has been extensively studied, mainly regarding carbon versus electrolyte optimization. For a better understanding of the phenomena occuring at the surface of the cathode, the charge transfer mechanism has been recently examined with either $SOCl_2$ or SO_2Cl_2 in neutral and acidic electrolyte [21],[22].

It has been shown that the reduction pathway is similar for the two products. In acidic electrolyte, the reactive species are rather onium ions like $[Cl_2Al(<-OSCl_2)_2]^+$ and $[Cl_2Al(SO_2Cl_2)]^+$. In neutral electrolyte, the reduction of species like $Li(SOCl_2)_2^+$ occurs. In every case, mass transport plays a role in the kinetics.

Under high rate conditions, the overall process appears to be greatly controlled by diffusion as demonstrated in $SOCL_2$ electrolyte [23] and in solvate based electrolyte containing SO_2Cl_2 [19]. In $SOCl_2$ electrolyte, it seems that the change from kinetic control to mass transport control occurs for current densities around 20 mA/cm^2 [25], [31].

To increase rate capabilities of the positive electrode, a large number of modifications have been proposed:

Use of high surface carbon black [24]
Use of pore formers [24]
Surface treatments [25], [26]
Addition of metal: [27]

5

Use of metallic electrocatalyst:[28], [29]

Addition of organometallic complexes:[30], [31], [32]

Addition of halogens to the electrolyte

Under high discharge rate, high surface carbons have been found to be preferable [24], and are actually used [38]. Nevertheless, Shawinigan type carbons are commonly satisfactory for high rate applications [33], [34]. The use of pore formers like NH_4HCO_3 with Shawinigan/platinum based cathode results in higher load potential and capacity ($J = 140$ mA/cm^2) [24]. Leaching the cathode with acetone gives improved capacity ($J = 50$ mA/cm^2) [25]. Addition of copper both increases load voltage and capacity [25]. Among the large number of electrocatalyst, platinum performed the best.

Organometallic complexes such as phtalocyanines, porphyrines or tetraazaanulenes of transition metals have been widely studied. Incorporation of these compounds to the cathode results in voltage and capacity improvements. Their catalytic activity is attributed to lowering of the activation energy of the charge transfert process. However, in a review paper published recently, Abraham [35] shows that catalytic activity of these compounds is not obvious. The increased capacity can be attributed to changes in cathode morphology and modification of the LiCl deposit.

Because of the mass transport limitations occuring under high discharge rate, the distribution of the catalysts in the cathode is important to prevent a rapid decrease of their usefulness. The heat treatments used during the preparation and the associated structutral modifications of the carbon collectors are certainly the key point of their activity.

Other modifications include the addition of iodine to $SOCl_2$ or bromine to SO_2Cl_2 electrolytes which result in increased preformances [36], [37].

Further reductions of cathodic polarization are believe to be found in the modification of the cathode structure (porosity, electrochemical surface area, catalyst) which can be adapted to the electrolyte (cathodic material, salt, conductivity).

PRACTICAL EXAMPLES

The topics developped above are illustrated by practical examples which are limited by a lack of details about the electrochemistry. Moreover, the exact utilization of the batteries is not always specified because of the classification of military projects.

"Missile type" applications

In 1982, Hiller presented results of a 91 volts multistack battery able to deliver a specific power of 2900W/kg during pulses of 140 to 200 mA/cm^2. No precision on the electrochemistry is given.

Using a slightly acidic $SOCl_2$ electrolyte ($LiAlCl_4$ 1.5 M, $AlCl_3$ 0.5 M) and a high surface area carbon, a battery containing a 76 volts section has been discharged for 50 minutes at an average rate of 12 mA/cm^2. The final temperature was 150°C. The use of this lithium battery allowed a weight saving of 50% with a 50 % performance margin compared to a AgO/Zn battery [38].

Based on same requirements, a comparative study of AgO/ZN , thermal system and Li/SOCl2 battery led to the following performances [40]:

	Specification	Li/SOCl$_2$	AgO/Zn	Thermal LiAl/FeS$_2$
Weight (Kg)	3.2	2.75	3.5	4.3
Volume (cm^3)	1640	1740	3100	1850
Energy densities Wh/kg Wh/dm^3	36 70.7	63 99	45 51	30 70
Wet stand life	T.B.D. hours	> 10 hours	3 hours	No
End of discharge case temperature	< 100°C	85°C	65°C	250°C
Storage life	> 10 years	> 10 years	> 10 years	> 15 years

These results demonstrate the advantages of lithium systems over others from a weight view point. In addition, a wet stand time exceeding 10 hours is possible without degradation of performance due to the use of $SOCl_2$, $LiAlCl_4$ 3 M containing additives. The cathodes were made of a blend of YS acetylene black (similar to Shawinigan acetylene black) and PTFE without additives. The discharge rate was 53 mA/cm^2.

7

A 20 volts battery has been studied based on the use of an $NaAlCl_4$, 3 SO_2 electrolyte [20]. Suitable for discharge temperature comprised between -30°C and +70°C, the electrolyte contains 4 M/l of SO_2Cl_2 as cathodic active material and Br_2 to "catalyze" the reduction. During the discharge, C to 67 C rates are used. The maximum current density is 150 mA/cm^2 and the corresponding voltages are 2.4 volts at -30°C and 3 volts at 70°C. These results were obtained on a volta pile design with YS acetylene black carbon.

Underwater applications

These applications depend on battery cooling by the electrolyte circulation throught a heat exchanger (cooled by sea water).

In addition, the electrolyte flow prevents rate limitations due to slow diffusion of reactive species. With $SOCl_2$ electrolyte, the circulation alleviates LiCl and sulfur deposition in the porous cathode [41]. With acidic electrolyte, discharge duration exceeding 16 minutes above 2.5 volts has been achieved under 200 mA/cm^2. Under 500 mA/cm^2, discharge duration is 2.5 minutes with a mean voltage of 2.7 volts [41].

With similar electrolyte, Eichinger [42] presented results at 100 mA/cm^2 with a discharge duration of 25 minutes above 3 volts. Tests of batteries revealed the possibility of reaching power output of 13.6 KW (52V, 262 A) under 200 mA/cm^2 without a catalyst. The flowing electrolyte provides for filtration to avoid sulfur and LiCl precipitation and for resaturation with $AlCl_3$.

Using a volta stack design of 117 cells and flowing SO_2Cl_2 electrolytes [43], 400 volt modules have been discharged during 18 minutes with a voltage comprised between 395 volts under 30 KW and 380 volts under 50 KW. In this case, the flowing electrolyte ensure thermal management of the system.

Pulsed applications

It has been found that pulse discharge capacity is improved over continuous discharge [9] using $SOCl_2$, $LiAlCl_4$ 1.8M and Shawinigan black cathodes. In some cases, the discharge is limited by anodic polarisation. Capacity maxima are obtained for optimal combinations of pulse width and rest time. The capacity improvements certainly results from diffusion of ionic species during rest time. Using 1.6 molar $LiGaCl_4$ solutions, current densities from 200 mA/cm^2 to 400 mA/cm^2 provide for voltages greater than 2 volts. Maximum power of 50 kW during pulses were achieved during the discharge of 20 cells stack [44]

FURTHER DEVELOPMENTS IN AMBIENT LIQUID CATHODE SYSTEMS

Rechargeable Li/SO$_2$Cl$_2$

In order to develop high rate rechargeable cells, the reversibility of the lithium/SO$_2$Cl$_2$ has been studied [45],[46]. The cathode efficiency is found to be quite good (up to 83 % [45]) and important improvements can be achieved by the used of solvate LiAlCl$_4$, 3 SO$_2$ based electrolyte [46]. Nevertheless, the lithium reversibility seems to be the weak point with dendrites causing cells shorting during cycling.

Substitute liquid cathodes

To find a "safer" and non toxic substitute to sufur oxychlorides, series of halocarbon have been tested [47],[48]. The highest discharge voltage (1.5 volts) under 1 mA/cm^2 was obtained with 1,1,2,2 tetrachloroethane (CHCl$_2$CHCl$_2$) added to a THF, LiAsF$_6$ electrolyte. Using catalysed doped carbon, a voltage of 2.3 volts at 1 mA/cm^2 was obtained. In these conditions, energy density should be 50% that of Li/SOCl$_2$ cells.

Looking for high rate reversible system, tests have been made with organosulfur compounds [49]. At 16 mA/cm^2, the discharge voltage is less than 2 volts. Despite the promising energy density, the output power of these systems seems to be too low, and they are therefore not competitive with sulfur oxychloride systems.

Bromine trifluoride batteries

Of the various interhallogen compounds, bromine trifluoride has the most attractive properties with the highest intrinsic conductivity due to a self ionization [50], [51], [52]. With lithium, the cell OCV (5.2 volts) approaches that of the "ultimate" lithium/fluorine system. The theoritical energy density of 2680 Wh/kg is nearly twice as much that of Li/SOCl$_2$ system. The discharge performances appear to be limited by anodic polarization resulting from a very passive LiF film. Using platinum for BrF$_3$ reduction, a load voltage above 4.5 volts for 12 min. at 20 mA/cm^2 is obtained [50]. Rates of 50 mA/cm^2 can also be supported. Discharge duration is increased by the addition of LiSbF$_6$ or LiPF$_6$ to the electrolyte.

The lithium/BrF$_3$ is to date the most promising system in terms of energy density and rate capablities to compete with lithium/sulfur oxychlorides systems. Corrosion of the lithium anode limits it to a reserve system. The very high reactivity of BrF_3 is the major limitation to wider use.

CONCLUSIONS

The high rate lithium/sulfur oxychlorides couples remain the most interesting ambient systems in terms of energy density and specific power, especially in reserve configuration.

Rate capabilities can be increased over a larger temperature range using SO_2 solvate based electrolytes. Better load voltage and capacity seem attainable by optimization of:

electrolyte composition

carbon electrode structure

Basic knowledge of reaction paths will lead to improvements.

Because of the highly specific requirements attached to the development of these systems, high energy batteries can be developped combining high reliability and safety. A good knowledge of thermal management and pressure variations during the discharge are the necessary conditions of the success of these batteries.

Among the various "new" liquid cathode systems presently under study, Li/BrF_3 is certainly the most promising.

ACKNOWLEDGEMENT

The authors wish to thank Dr M. Broussely and S. Hafner for helpfull discussions and revision of this paper.

REFERENCES

[1]   H.F. GIBBARD, 4th international Meeting on Lithium Batteries, May 24-27, 1988, Vancouver B.C., Canada.

[2]   D.L. CHUA, W.J. EPPLEY, R.J. HORNING (chapter 39) in Handboook of Batteries and Fuel Cells , D. LINDEN Editor, McGraw Hill.

[3]   S. EBEL, M. ZELINSKY, P. SIZE, Proceedings of the 33rd International Power Sources Symposium, The Electrochemical Society, Pennington, NJ, p156, (1988).

[4]   C.R. SCHLAIKJER in "Lithium Batteries" (J.P. GABANO Ed.) Academic Press (1983).

[5]   D. VALLIN and P. CHENEBAULT, Proceedings of the 32nd International Power Sources Symposium, 9-12 june, 1986, Cherry Hill, NJ.

[6] Y.I. CHO, H. FRANK and G. HALPERT, Journal of Power Sources, 21 (1987) 183-194.

[7] K.A. KLINEDINST, Proceedings of the 31st International Power Sources Symposium, Atlantic City, NJ, 1982.

[8] V. DANEL, J.P. DESCROIX, G. SARRE, Proceedings of the 32nd International Power Sources Symposium, 9-12 june, 1986, Cherry Hill, NJ.

[9] R.C. McDONALD, P. HARRIS, F. GOEBEL, Proceedings of the 33rd International Power Sources Symposium, 1988, Cherry Hill, NJ.

[10] A.J. HILLS and N.A. HAMPSON, Journal of Power Sources, 24 (1988) 253-271.

[11] G.E. BLOMGREN, Proceedings of the Symposium on Lithium Batteries, Vol 87-1, A.N. DEY Ed., the Electrochemical Society Inc, Pennington, NJ.

[12] W.P. HAGAN, N.A. HAMPSON, R.P. PACKER, Journal of Power Sources, 24 (1988) 95-113.

[13] S.D. JAMES, Proc. of the Symposium on Lithium Batteries, 1983, Vol 84.1, A.N. DEY Ed., the Electrochemical Society Inc, Pennington, NJ.

[14] M. DOMENICONI et Coll, (1977) Interim Report, Contract N° N00014-76-C-0524.

[15] K.A. KLINEDINST, NOSC Contract, N° N000123-81-C-443.

[16] Y. BEDFER, J. CORSET, Journal of Power Sources, 9 (1988) 267-272.

[17] P.A. MOSIER-BOSS, S. SZPAK, J.J. SMITH, R.J. NOWAK, J. Electrochemical Society, 136,1282 (1989).

[18] R.J. MAMMONE, M. BINDER, J. Electrochemical Society, 134, 1 (1987).

[19] E. LOJOU, R. MESSINA, J. PERRICHON, J.P. DESCROIX, G. SARRE, The Electrochemical Society, Fall Meeting, (1987) Oct 18-23, Honolulu, HAWAII.

[20] P. CHENEBAULT, J.P. PLANCHAT, E. LOJOU, Proceedings of the 34th International Power Sources Symposium, 1990, Cherry Hill, NJ, IEEE Service Center, Piscataway, NJ.

[21] P.A. MOSIER-BOSS, S. SZPAK, J.J. SMITH, R.J. NOWAK, Electrochimica Acta, Vol. 35, N°11/12, pp1787-1795, 1990.

[22] P.A. MOSIER-BOSS, S. SZPAK, R.J. NOWAK, J.J. SMITH, in Power Sources 13, Proceedings of the 17th International Power Sources Symposium held in Bournemouth, UK, April 1991, T. KEILLY and B.W. BAXTER Ed., International Power Sources Symposium Committee.

[23] W.P. HAGAN, N.A. HAMPSON, R.P. PACKER, Electrochimica Acte Vol.31, N°6, pp 699-704, 1986.

[24] K.A. KLINEDINST, J. Electrochemical Society, 132, N°9, 2044, (1985).

[25] W.P. HAGAN, N.A. HAMPSON, R.P. PACKER, Power Sources 11, 1987, p 413, L.J. PEARCE Ed, Taylor and Francis Ltd, Basingstoke, UK

[26] C.W. WALKER, M. BINDER, W.L. WADE, S. GILMAN, J. Electrochemical Society, 132, 1536 (1985).

[27] W.K. BEHL, Proceedings of the Symposium on Lithium Batteries, Vol 81-4, H.V. VENKATASETTY Ed., the Electrochemical Society Inc, Pennington, NJ.

[28] K.A. KLINEDINST, J. Electrochemical Society, Vol.128, 2507 (1981).

[29] C.Y. OH, Y.Y. WANG, C.C. WAN, Journal of Power Sources, 16 (1985) 233.

[30] N. DODDAPANENI, Proceedings of the Symposium on Lithium Batteries, Vol 84-1, A.N. DEY Ed., the Electrochemical Society Inc, Pennington, NJ.

[31] W.P. KILROY, M. ALAMGIR, S.J. PERROTTI, K.M. ABRAHAM, The Electrochemical Society, Fall Meeting, (1987) Oct 18-23, Honolulu, HAWAII.

[32] K.M. ABRAHAM, L. PITTS, W.P. KILROY, J. Electrochemical Society, Vol.132, 2301 (1985).

[33] V. DANEL, J.P. DESCROIX, A. PETIT, Proceedings of the Symposium on Lithium Batteries, Vol 84-1, A.N. DEY Ed., the Electrochemical Society Inc, Pennington, NJ.

[34] P. HARRIS, G. BOYLE, F. GOEBEL, Proceedings of the 32nd International Power Sources Symposium, 9-12 june, 1986, Cherry Hill, NJ.

[35] K.M. ABRAHAM, Journal of Power Sources, 34, (1991) 81-101.

[36] K.A. KLINEDINST, J. Electrochemical Society, Vol.137, N°2, 398 (1990).

[37] V. DANEL, J.P. DESCROIX, G. SARRE, Proceedings of the 32nd International Power Sources Symposium, 9-12 june, 1986, Cherry Hill, NJ.

[38] M. PEABODY, R.A. BROWN, Proceedings of the 33rd International Power Sources Symposium, The Electrochemical Society, Pennington, NJ, p312, (1988).

[39] R. HILLER, J.R. DRISCOLL, Proceedings of the '30th Power Sources Symposium, Atlantic City, Nj, june 7-12, 1982, The Electrochemical Soceity, p198.

[40] J.P. PLANCHAT, J.P. DESCROIX, G. SARRE, Proceedings of the 33rd International Power Sources Symposium, The Electrochemical Society, Pennington, NJ, p312, (1988).

[41] M. PUSKAR, P. HARRIS, Proceedings of the 32nd International Power Sources Symposium, 9-12 june, 1986, Cherry Hill, NJ.

[42] G. EICHINGER, G. SEMRAU, W. JACOBI, J. HEYDECKE, Proceedings of the 34th International Power Sources Symposium, 1990, Cherry Hill, NJ, IEEE Service Center, Piscataway, NJ.

[43] M.P. LANOT, L. D'USSEL, J. HASTINGS, Proceedings of the 34th International Power Sources Symposium, 1990, Cherry Hill, NJ, IEEE Service Center, Piscataway, NJ.

[44] P. HARRIS, M. GUENTERT, F. GOEBEL, Proceedings of the 34th International Power Sources Symposium, 1990, Cherry Hill, NJ, IEEE Service Center, Piscataway, NJ.

[45] P.H. SMITH, S.D. JAMES, J. Electrochem. Soc., Vol 137, (1990), 602.

[46] E. LOJOU, R. MESSINA, P. CHENEBAULT, G. CREPY, B. SIMON, to be published in the J. Electrochem. Soc.

[47] P.H. SMITH, S.D. JAMES, K.M. O'NEIL, M.H. WILSON, J. Electrochem. Soc., Vol 136, (1989), 1625.

[48] P.H. SMITH, S.D. JAMES, K.M. O'NEIL, M.H. WILSON, J. Electrochem. Soc., Vol 136, (1989), 1631.

[49] S.J. VISCO, M.LIU, L.C. DE JONGHE, J. Electrochem. Soc., Vol 137, (1990), 1191.

[50] M.H. MILES, K.H PARK, Proceedings of the 34th International Power Sources Symposium, 1990, Cherry Hill, NJ, IEEE Service Center, Piscataway, NJ.

[51] M.H. MILES, K.H PARK, D.E. STILWELL, J. Electrochem. Soc., Vol 137, (1990), 3393.

[52] G.L. HOLLECK, G.S. JONES, Proceedings of the Symposium on Primary and secondary Lithium Batteries, Vol 91-3, K.M. ABRAHAM, M. SALOMON Ed., the Electrochemical Society Inc, Pennington, NJ.

DESIGN ANALYSIS OF BIPOLAR Li-TiS$_2$ BATTERIES

D. H. Shen, A. Attia, S. Subbarao, and G. Halpert

Jet Propulsion Laboratory
California Institute of Technology
4800 Oak Grove Drive
Pasadena, California 91109

ABTRACT

The present study uses an empirical model to assess the feasibility of using the Li-TiS$_2$ bipolar battery for high power applications. Predicted performance outputs at a variety of conditions were calculated. The effects of the design parameters on the performance of bipolar Li-TiS$_2$ batteries are presented. Specific energies greater than 150 Wh/kg can be achieved at low rates. Specific power levels in excess of 100 W/kg can be reached at high rates but with a reduction of the specific energy to less than 70 Wh/kg.

INTRODUCTION

In recent years, many defense and transportation pulse power applications [1] are emerging, that require power sources with high specific power, high specific energy, and long cycle life. Further, these applications require power systems to operate in pulse mode for longer duration on the order of seconds to minutes. At present, capacitors are used for pulse power applications; however, they cannot satisfy the long duration pulse requirement. Conventionally, batteries are used for applications requiring low power and high capacity, but the development of bipolar designs can extended their use to high pulse power applications requiring seconds to minutes operation. Lead-acid bipolar batteries are under development to meet the high power requirements [2,3,4]. However, the low specific energy of lead acid battery is a serious draw back. A battery that provides both high specific energy and high specific power would be attractive for transportation and defense applications.

Lithium batteries have high specific energy (100 Wh/kg) compared to other rechargeable battery systems (30-50 Wh/kg). Small capacity rechargeable lithium cells of monopolar design such as Li-TiS$_2$, Li-CoO$_2$, Li-MoS$_2$, Li-NbSe$_3$, Li-V$_3$O$_8$, Li-V$_2$O$_5$, etc. are currently being made. Bipolar rechargeable lithium batteries would be attractive because of their potential for high specific energy as well as high specific power. In this

13

paper, an attempt is made to study the design options for bipolar lithium batteries. The Li-TiS$_2$ system was selected in this study because of its long cycle life capability [5] and proven performance data base. Further, it was reported [6] that thin, oriented TiS$_2$ microcrystallite films were capable of delivering high current pulses. A computer program was developed to examine the influence of the design parameters on battery performance. Details of the program and the projected performance capabilities of bipolar rechargeable Li-TiS$_2$ batteries are reported.

DESIGN PROGRAM

A computer program was developed to evaluate the effects of various design options on the performance of bipolar Li-TiS$_2$ batteries. This program will generate a battery design based on desired design inputs and component design data base built into the program. Major steps in the execution of the battery design computer program are: 1) determination of the actual cathode and anode capacity based on the component performance data base, 2) optimization of bipolar plate dimension, 3) determination of number of cells required based on battery voltage and discharge rate, 4) calculation of quantity of electrolyte required, and 5) design definition of other components.

The design inputs for the program are: battery capacity, battery voltage, and discharge rate. The component design data base consists of design parameters (variables and constants) and component performance characteristics based on experimental results. Some of the important design parameters considered are electrode characteristics (dimensions and utilization), current densities, separator dimensions, electrolyte and separator properties, and material properties of the battery hardware. Table 1 summarizes the important design parameters that have been considered in the present study. Electrode dimensions and current densities are the main variables in the program along with the design inputs. Design parameters such as anode to cathode capacity ratio, electrode height to width ratio, thickness of the bipolar substrate, can wall thickness, number of separator layer, thickness of end plate, and overrating factor were kept constant in this study. Different values for these design constants can be specified by the user if required. Cathode utilization and average discharge voltage at different current densities (Figure 1) are established from in-house experimental results using 15 mil thick TiS$_2$ cathodes at 25°C. The program details are summarized in Table II. Some major issues that have not been considered in the study are: thermal design parameters, temperature effect on cathode utilization, and the degradation of the electrolyte.

14

RESULTS

The impact of battery capacity, battery voltage, and discharge current on specific energy (Wh/kg) was determined using this design program. Designs for two medium rate (C/2) batteries and their projected performance were also determined.

Influence of Battery Capacity on Specific Energy

The effect of battery capacity (1 to 250 Ah) on the specific energy of a 50 V battery as a function of battery capacity at C/10 and C/2 rates is given in Figure 2. From the figure, it can be observed that battery specific energy increases marginally (20-30%) with increase in capacity. The specific energy was found to increase up to 10 Ah and level off thereafter. At low battery capacities, the nonactive components contribute a significant percentage to the weight of the total battery, which results in low specific energy. As the battery capacity increases beyond 10 Ah, the nonactive material weight contribution in reduced resulting in improved specific energy. However, the specific energy does not increase significantly beyond 10 Ah because the battery configuration becomes plate-like shaped, offsetting the gain from the reduced weight of the inactive components.

Influence of Battery Voltage on Specific Energy

The effect of battery voltage (5 to 250 V) on the specific energy of a 10 Ah battery at C/2 and C/10 discharge rates is shown in Figure 3. From the figure, it can be observed that the specific energy is very sensitive to the bipolar battery voltage. The specific energy of the batteries increases significantly up to 100 V and level off afterwards. The gain in specific energy with battery voltage is mainly due to the reduction in the percentage of battery hardware weight contribution to the overall battery weight. Higher voltage batteries are cubic in shape compared to lower voltage batteries which are plate-like in configuration; hence, the high voltage batteries require less battery casing per unit Wh. The bipolar configuration is not attractive for batteries with voltages less than 10 V.

Influence of Discharge Rate on Specific Energy

The effect of battery discharge rate on the specific energy of two 10 Ah batteries (50 and 250 V) is given in Figure 4. Various battery discharge rates, from C/20 to 2C,

15

were studied. The specific energy is found to be sensitive to the discharge rate of the battery. A specific energy greater than 150 Wh/kg can be achieved at low discharge rate. However, the specific energy drops by more than 50 to 100 Wh at rates higher than 1C. The decrease of the specific energy of the batteries is due to poor electrolyte conductivity and low TiS_2 utilization at high discharge rate. Use of improved electrolytes and optimization of the bipolar plate structure may improve the rate capability of the bipolar $Li-TiS_2$ battery.

Design Feasures and Performance Characteristics of Two Typical $Li-TiS_2$ Bipolar Batteries

The outputs of the program contain battery design features and projected performance characteristics (specific energy, energy density, and specific power). The weight budget of various components of these two batteries are given in Figure 5. It can be observed that 50% of the weight of the 10Ah, 50V medium rate (C/2) battery is contributed by the hardware and current collector. However, the contribution from the battery hardware and current collector decreased to ~30% in a 50Ah, 250V medium rate (C/2) battery and due to this reason, the specific energy of this battery is greater by ~30% compared to the 10Ah, 50V medium rate (C/2) battery. Ragone plots of these two batteries are given in Figure 6. The specific energy of the batteries is found to decrease with increasing specific power. This finding is in agreement with the discharge rate effects as discussed above. In summary, a bipolar $Li-TiS_2$ battery using 15 mil thick TiS_2 bipolar plate can be operated as a medium power and high specific energy battery.

Battery service life at various discharge loads for a 10Ah, 50V battery is shown in Figure 7. The amperes per kilogram vs. hours of service log-log plot can be related to Peukert's equation [8],

$$I^n t = C$$

where I is the discharge rate and t is the corresponding discharge time, and can be used to describe the performance of a battery. The value n is the slope of the straight line and is approximately equal to 1. The curve is linear on this log-log plot of discharge load vs. discharge time which follows Peukert's equation. The curve tapers off at high current indicating that this type of battery is not able to operate at high current. There is no tapering off at the low current end because of the low self discharge rate of the battery.

CONCLUSIONS

Some of the important findings of a Li-TiS$_2$ bipolar battery design analyses are the following: 1) specific energy greater than 150 Wh/kg can be achieved for Li-TiS$_2$ bipolar batteries; 2) bipolar lithium batteries can be operated at 2C rate continuously and the realizable specific energy and power are ⁻50 Wh/kg and ⁻100 W/kg, respectively; 3) bipolar designs are not attractive for battery capacities lower than 10 Ah and voltages less than 10 V; 4) the battery specific energy increases only marginally (20-30%) with increase in capacity and levels off after 10Ah; and 5) the specific energy is found to be sensitive to the discharge rate. In summary, a bipolar Li-TiS$_2$ battery using 15 mil thick TiS$_2$ bipolar plate can be operated as a medium power and high specific energy battery.

ACKNOWLEDGEMENT

The work described here was carried out under contract with the National Aeronautics and Space Administration (Code R).

REFERENCES

1. Burke, A.F., Hardin, J.E., and Dowgiallo, E.J. "Application of Ultracapacitors in Electric Vehicle Propulsion Systems", _Proceedings of Power Source Symposium_, Cherry Hill, May 1990.

2. Rowlette, J.J. "Optimized Design Variables for High Power Batteries", _IECEC_, 1987

3. Kao, W.H. "Computer Aided Design of a Bipolar Lead-Acid Battery", _ECS Extended Abstract_, Fall 1990.

4. Attia, A, Perrone, D.E. "Sealed Bipolar Lead-Acid Batteries for High Pulse Power Applications", _Symposium of International Lead Zinc Reserach Organization_, August 1989

5. Shen, D.H., Surampudi, S., Huang, C.-K., Deligiannis, F., Halpert, G., Dominey, L., Koch, V.R., and Goldman, J., "Improved Lithium Electrode Cycling in Ether Based Electrolytes with Synergistic Additives", Part I. Microcalorimetry, AC Impedance Spectroscopy and Full Cell Cycling Studies, _Proceedings of Fall ECS Meeting_, Seattle, Washington, Oct 14-19, 1990

6. Koch, V.R., "Performance of Bipolar Li/TiS$_2$ Pulse power Batteries", <u>The 5th International Lithium Batteries Seminar</u>, Florida, March 1991

7. Linden, D., "Handbood of Batteries and Fuel Cells", McGraw-Hill, 1984

TABLE I. DESIGN PARAMETERS OF BIPOLAR Li-TiS$_2$ BATTERY

COMPONENTS	DESIGN PARAMETERS
CATHODE	THICKNESS; POROSITY; MATERIAL UTILIZATION; WIDTH; HEIGHT TO WIDTH RATIO
ANODE	NEGATIVE TO POSITIVE CAPACITY RATIO; LITHIUM FOIL THICKNESS; UTILIZATION
ELECTROLYTE	COMPOSITION; DENSITY; QUANTITY
SEPARATOR	TYPE; THICKNESS; POROSITY; DENSITY; NO. OF LAYERS
BIPOLAR SUBSTRATE	DENSITY; THICKNESS; CONDUCTIVITY
FRAME	MATERIAL; DIMENSION; DENSITY
END PLATE	AREA; THICKNESS; DENSITY
CASE	WALL THICKNESS; DENSITY; SEAL; OVERHEAD SPACE

TABLE II. OVERVIEW OF COMPUTER PROGRAM FOR BIPOLAR BATTERY DESIGN

INPUTS	DESIGN CONSTANTS	DATA BASE	OUTPUTS
*DESIGN CAPACITY	*ANODE TO CATHODE CAPACITY RATIO: 3 TO 1	*CATHODE UTILIZATION VS. CURRENT DENSITY	*SPECIFIC ENERGY
*BATTERY VOLTAGE	*HT TO WIDTH RATIO: 1 TO 1	*AVERAGE DIS. VOLTAGE VS. CURRENT DENSITY	*ENERGY DENSITY
*DISCHARGE RATE	*CATHODE THICKNESS: 15 MIL		*SPECIFIC POWER
	*CURRENT COLLECTOR: S.S. 1 MIL		*ACTUAL CAPACITY
	*CAN: S.S. 20 MIL		*ACTUAL VOLTAGE
	*SEPARATOR: CELGARD 2400; 2 LAYERS		*CELLS # REQUIRED
	*END PLATE: S.S. 50 MIL		*COMPONENT DIMENSIONS
	*OVER RATING FACTOR		*COMPONENTS WT%

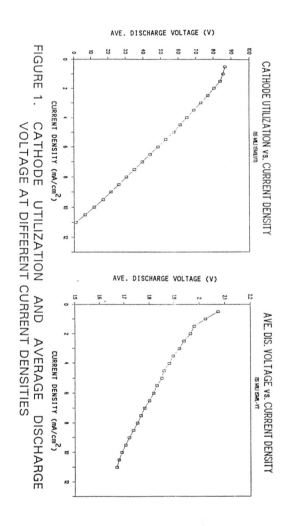

FIGURE 1. CATHODE UTILIZATION AND AVERAGE DISCHARGE VOLTAGE AT DIFFERENT CURRENT DENSITIES

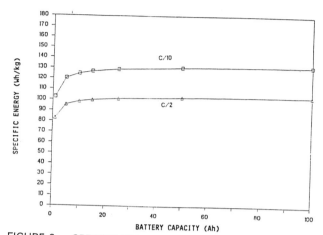

FIGURE 2.   SPECIFIC ENERGY OF A 50 V BATTERY AS A FUNCTION OF
BATTERY CAPACITY AT C/10 AND C/2 RATES

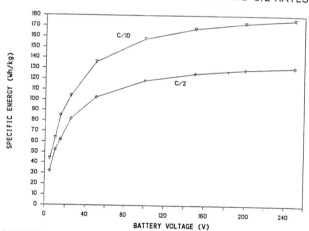

FIGURE 3.   THE EFFECT OF BATTERY VOLTAGE (5 TO 250 V) ON
SPECIFIC ENERGY OF A 10 Ah BATTERY AT C/2 AND C/10
DISCHARGE RATES

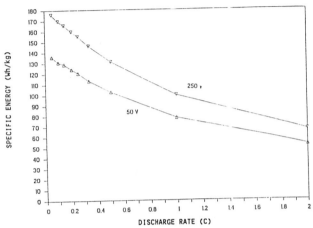

FIGURE 4.   THE EFFECT OF BATTERY DISCHARGE RATE ON THE
SPECIFIC ENERGY OF TWO 10 AH BATTERIES(50 & 250 V)

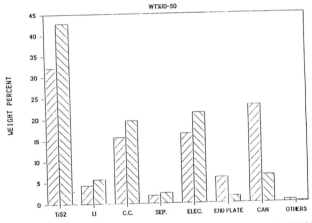

FIGURE 5.   THE WEIGHT BUDGET OF VARIOUS COMPONENTS OF
10Ah, 50V AND 50AH, 250V BATTERIES

22

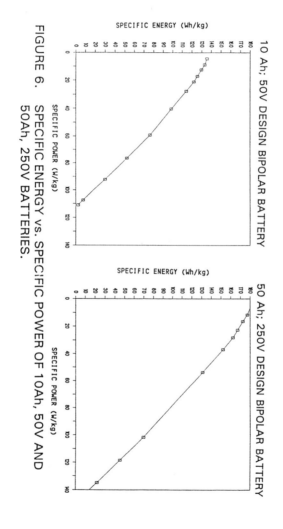

FIGURE 6.   SPECIFIC ENERGY vs. SPECIFIC POWER OF 10Ah, 50V AND
50Ah, 250V BATTERIES.

23

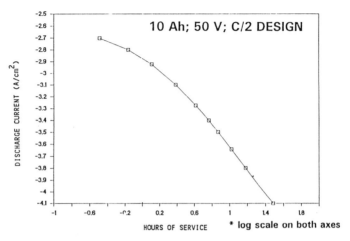

FIGURE 7.   BATTERY SERVICE LIFE AT VARIOUS DISCHARGE
LOADS FOR A 10Ah, 50V BATTERY

24

Performance and Safety Features of Spiral Wound Lithium / Thionyl Chloride Cells

G. Eichinger and W. Gabriel

Sonnenschein Lithium GmbH, Industriestrasse

D-6470 Büdingen, Germany

Abstract

The design of spiral wound lithium / thionyl chloride cells is discussed. Special emphasis is put on the features which influence safety. Performance data of such cells over a variety of discharge temperatures and discharge currents are given. The results of abuse tests like charging and forced overdischarge are presented. The cells show high capacities and behave well under all conditions.

Introduction

From the fairly large number of lithium battery systems, which have been investigated since the early seventies, only few found their way into mass production. One of these is the lithium / thionyl chloride system, which shows excellent energy density data (gravimetric as well as volumetric). In many high-rate applications, especially for man-pack radios, however, lithium/sulfur dioxide is still the favourite system. For such applications, lithium/thionyl chloride is still in the start-up phase. In this paper, performance and safety features of high-rate lithium / thionyl chloride cells, especially of the C-size type, are described.

Experimental

Special emphasis was put on the testing of the vent. The vent is stamped into the bottom of the can and has a cross shape. By applying suitable tooling and a stringent quality control, it was possible to obtain a high reliable vent.
The pressure at which the vent operates was tested by using a test set-up working with hydraulic oil pressure. The increase of pressure during the operation of this test fixture and the sudden decrease after rupture of the vent, was recorded with an oscilloscope.
To get a suitable signal for recording, a pressure transducer (Kistler, type 7011) was used.

After the design of the vent was finished by fixing the remaining wall thickness of the can at the vent after the stamping process, the function of the vent was tested in complete cells.

For this purpose, test cells were put in an oven and heated up to 90°C. After 2 h at this temperature, the cells were controlled for venting and the temperature was increased in steps of 10°C between the a.m. control steps (cells remained each time 2 h at the adjusted temperature). The vent operates very well and opens at temperatures between 120°C-140°C.

All test cells were prepared according to development specifications in a dry room at a relative humidity below 2 %. All electrical tests were made at constant current.

For short circuit protection, single cells are equipped with a reversible fuse, limiting the maximum current at room temperature to 2 A. The applied polyswitch is integrated in the can / cover assembly.

Fig. 1 gives a cross-sectional view through the cell.

Results

Design features of the spiral wound C cell

The can has a stamped vent in the bottom having a cross shape. The operating pressure of this vent was adjusted to 40 ± 3 bar. Whereas in the first test samples the stamping process was made by hand, later on it was made in a production scale as an additional step in the line after deep drawing the can. A very careful quality control is made to ensure a good accuracy in the operation of the vent.

For a proper function besides the mechanical parameters of the vent itself, the void volume in the cell after filling of the electrolyte has to be adjusted within small tolerances. This can be achieved only by having narrow tolerances in the core of the cell and by filling a controlled amount of electrolyte. The reason is, that in fresh cells, the vent operates obviously not due to gas overpressure inside the cell, but more due to hydraulic pressure of the electrolyte.

This situation changes in a partially or completely discharged cell, where due to the discharge reaction, sulfur dioxide is present in the cell and electrolyte is consumed according to the electrochemical discharge reaction. At room temperature the sulfur dioxide formed, is obviously bound in complex form with the $LiAlCl_4$, since only a very small overpressure can be found in discharged cells. At high temperatures, these complexes decompose, thus forming sulfur dioxide overpressure.

A large number of publications (e.g. 1-4) deals with thermal modelling and heat dissipation in spiral wound lithium cells. These calculations, however, seem to be important only in the case of very high rate discharge. At mean discharge rates, normally the heat transfer in the cell is not a major problem.

To be on the safe side, regarding this heat transfer, the cell shall not be cathode (electrolyte) limited, since in that case, the internal resistance at the end of discharge might increase to very high values. This can happen especially at constant current discharge thus causing overheating of the cell

at the end of the discharge or at the beginning of a forced overdischarge
Lithium limited cells, however, produce considerably less heat at the end of
discharge than cathode limited cells (5).
Due to these facts the cells were designed to be lithium limited with excess
of electrolyte. For a safe spiral wound cell, another important point is, that
the lithium is consumed uniformely during discharge and that no electrically
insulated islands of lithium are left at the end of discharge. Especially at
the end of a forced high rate discharge, hot spots could be formed, if such a
situation occurs. To avoid this possibility, lithium is pressed on an expanded
metal screen.

Performance data of spiral wound C cells (Li/SOCl$_2$):

Regular Discharge

The cells show excellent discharge data up to a continuous discharge current
of 1 A. In fact higher discharge currents are also possible, but the testing
was mainly focussed on the range between 20 mA to 1 A discharge current.
Typical discharge curves at 20 mA and at 200 mA at ambient temperature
are shown in fig. 2.
Fig. 3 shows the correlation between capacity and discharge current.
As expected, the available capacity decreases significantly with increasing
discharge current.
However, even at 200 mA at room temperature, about 4.5 Ah capacity are
delivered without problems. A typical Li/SO$_2$ cell exhibits only 3.2 - 3.4 Ah
under similar discharge conditions.
The correlation between capacity and temperature is shown in fig. 4 for two
discharge currents (20 mA and 200 mA).
As expected, at high currents the capacity at low temperatures decreases
more, than at low currents. It should be notified, however, that the electrolyte was 1.5 M LiAlCl$_4$ and thus not optimized for low temperatures. An
overview on the discharge results which were observed under different conditions (currents and temperatures) is given in table 1. Besides the regular
discharge tests, the cells were subjected to environmental tests, like shock
and vibration according to MIL specifications. The discharge results were
not influenced by these preceding environmental tests. The cells exhibit a
very stable voltage plateau during discharge. The mean voltage of this
plateau in dependence of the discharge current is given in fig. 5. Even at
the high discharge current of 1 A, the mean voltage during discharge at
R.T. is above 3.1 V.
After 4 weeks storage at 70°C, the cells showed no performance degradation
in capacity at a discharge with 20 mA at 15°C. The available capacity was
between 5.97 Ah and 6.15 Ah (at 2.4 cut-off) and thus comparable to the
results obtained with fresh cells under the same conditions (5.99 Ah and 6.36
Ah).
The voltage delay after this temperature storage was also no problem. The
voltage was recorded with a high speed voltmeter (HP 3437A) at a measuring frequency of 2000 points /s.
The lowest voltage at the a.m. discharge current was at -30°C above 3.4 V
(after a few millseconds) and had a comparable value at +60°C. At +15°C,
the lowest voltage was 3.6 V.

At a discharge rate of 200 mA, also no degradation in capacity after a 4 week storage at 70°C was observed.

Abuse Tests

Forced Overdischarge

Especially for battery packs, it is of utmost importance that a cell is tolerant to abuse, especially to voltage reversal.
For this situation, one has to be aware of two different mechanisms. Cells can be anode limited, though enough lithium is still present in the cell. This might be due to a bad contact between lithium and the current collector or to a blocking of the active lithium surface. If the cell is designed and built properly, however, it is more likely, that the cell becomes cathode limited due to plugging of the cathode pores. Depending on the discharge rate, in the same cell type, one can find the situation that the cell goes to voltage reversal (under forced discharge conditions at relatively high currents, e.g. 500 mA), though approximately 40-50 % of the lithium are still present in the cell. Furthermore, it was checked that this lithium is still in good contact with the metallic current collector. Since it is relatively unlikely that the lithium surface is blocked by insulating products, this is the situation of a cathode limited cell. This cathode limitation is due to a blocking of carbon substrate since enough electrolyte is left. After voltage reversal, the dominant reaction at the carbon cathode will be the deposition of lithium (6, 7). The situation which is established under these conditions is obviously not critical. All tests that were made with cells driven into voltage reversal at high currents showed a completely safe behaviour of the cells (table 2).
If the discharge current, however, is low (only a few mA) a completely different situation may arise, though cells of the same design are used for the test. At such low discharge currents, the cells are really anode limited, since all lithium is consumed before the voltage reversal occurs.
The reactions which may occur in this situation are described in l.c. 7, 8.
Reaction products found in such cells with IR and partly also with cyclic voltammetry include Cl_2, SCl_2, SO_2Cl_2 and $SOCl^+AlCl_4^-$ (8). SO_2Cl_2 seems to be a reaction product from SO_2 (regular discharge product) and Cl_2 (formed during forced overdischarge).
The formation of ClO_2 and Cl_2O, which was proposed earlier (9), however, was not verified in later work (5, 6).

Charging of the cells

For a further specific abuse test, cells were charged with 500 mA at various temperatures. For this experiment completely discharged and half discharged cells were used. The results are presented in table II.
Only the charging with 500 mA of the completely discharged cells at room temperature lead to a venting of two out of three test cells.
In all other cases no venting occured.
During charging, the cell voltage adjusts at a plateau of about 3,8 - 4 V. Fig. 6 shows this effect.

28

In all these experiments, the cells were charged or put on forced overdischarge for a time sufficient to charge or overdischarge the cells with a capacity of about 7.5 Ah. This is more than the maximum available capacity from the cells.

Conclusions

Spiral wound lithium/thionyl chloride cells with an anode limited design were developed and tested very thoroughly. It was proven by a large number of regular discharge tests and by abuse tests, especially charging and forced overdischarge that the system is safe. Potential safety risks seem to be associated much more with mechanical problems, like e.g. insufficient contact between lithium and the current collector in the course of discharge, than with the chemistry of the system.

Provided that the cells are produced in a reproducible and reliable way with narrow tolerances for the cell components, there is no reason why the lithium/thionyl chloride system should not be comparably safe as other lithium high rate systems.

REFERENCES

1. T.I. Evans and R.E. White, J. Electrochem. Soc. 136, 2145 (1989)

2. Y.I. Cho and G. Halpert, J. Power Sources 18, (1986) 109

3. Y.I. Cho, J. Electrochem. Soc. 134, 771 (1987)

4. Y.I. Cho and Dong Wook Chee, ibid. 138, 927 (1991)

5. K.M. Abraham, L. Pitts and W.P. Kilroy, ibid. 132, 2301 (1985)

6. B.J. Carter, H.A. Frank and S. Szpak, J. Power Sources 13, 287 (1984)

7. K.M. Abraham, R.M. Mank and G.L. Holleck, in Proc. Symp. Power Sources for Biomed. Implant. Appl. and Ambient Temp. Lithium Batteries (B.B. Owens and N. Margalit Eds), The Electrochem. Soc. Inc., Princeton N.J., 1980

8. A.I. Attia, Ch. Sarrazin, K.A. Gabriel and R.P. Burns, J. Electrochem. Soc. 131, 2523 (1984)

9. D.J. Salmon, M.E. Petersen, L.L. Henricks, L.L. Abels and J.C. Hall, ibid. 129, 2496 (1982)

Tab. 1
Typical Discharge Results

Discharge Current	Temperature	Capacity (cut-off 2.5V)
mA	°C	Ah
20	+63	5.9
	RT	6.0
	-30	4.1
100	+63	6.5
	RT	4.6
	-30	2.5
200	+63	6.2
	RT	4.3
	-30	2.0
500	+63	5.4
	RT	3.4
	-30	2.1
1000	+63	4.2
	RT	2.4
	-30	1.8

Tab. 2
Abuse Tests

Test Condition	Temperature °C	Current mA	Status of Cell	Effect
Charging	+60	500 mA	Disch. to	none
	RT	at 10V	50% of	none
	-30	for 15h	Capacity	none
Charging	+60	500 mA	Completely	none
	RT	at 10V	discharged	vent open
	-30	for 15h	(to 0V)	none
Forced Over-discharge	+65	1000 mA	After	none
	RT	more than	complete	none
	-40	7.5 h	discharge	none
	+65	500 mA	After	none
	RT	more than	complete	none
	-40	15 h	discharge	none
	+65	200 mA	After	none
	RT	more than	complete	none
		37.5 h	discharge	

30

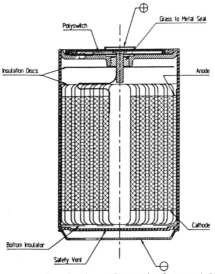

Polyswitch

Glass to Metal Seal

Insulation Discs

Anode

Cathode

Bottom Insulator

Safety Vent

Fig. 1: Cross sectional view of a spiral wound C cell (lithium/ thionyl chloride).

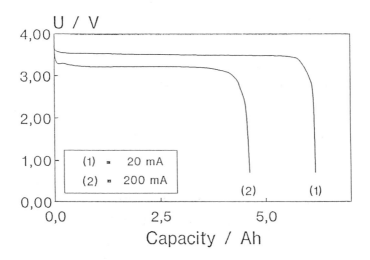

U / V

4,00

3,00

2,00

1,00

0,00

(1) = 20 mA
(2) = 200 mA

(2)   (1)

0,0   2,5   5,0

Capacity / Ah

Fig. 2: Typical discharge curves at 20 mA and at 200 mA constant current (spiral wound C cell, lithium/ thionyl chloride).

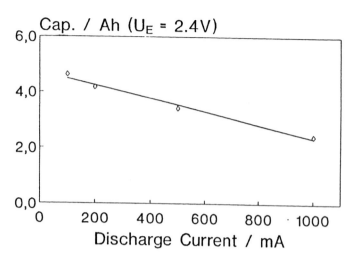

Fig. 3: Correlation between available capacity and discharge current at R.T. (spiral wound C cell, lithium/ thionyl chloride)

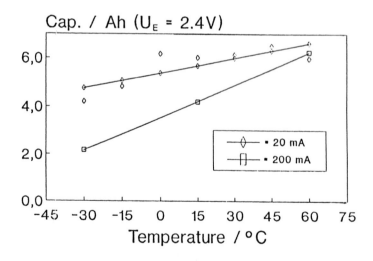

Fig. 4: Correlation between available capacity and discharge temperature at 20 mA and at 200 mA constant current (spiral wound C cell, lithium/ thionyl chloride).

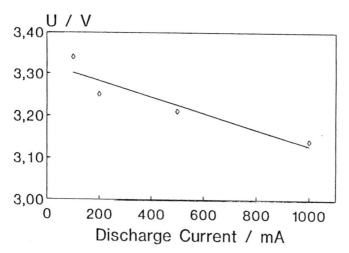

Fig. 5: Mean plateau voltage of the discharge curve at various discharge
currents (spiral wound C cell, lithium/ thionyl cloride).

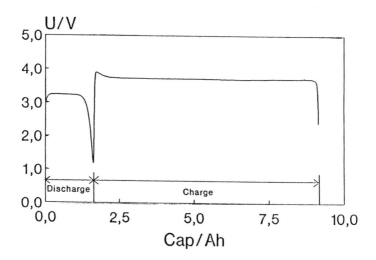

Fig. 6: Charging of a spiral wound C cell after a discharge of
approximately 50 % at a current of 500 mA (spiral wound C cell,
lithium/ thionyl chloride)

EXPOSURE OF MODERATE RATE LITHIUM/THIONYL CHLORIDE CELLS TO ELEVATED TEMPERATURE

Esther S. Takeuchi and Sally Ann Smesko

Wilson Greatbatch Ltd.
10,000 Wehrle Drive
Clarence, New York 14031

ABSTRACT

The effect of high temperature exposure (130°C) on lithium/thionyl chloride cells and cell components was examined in an attempt to identify the cause(s) of the decreased capacity and increased impedance normally exhibited by autoclaved cells upon discharge. The long term influence of heat treatment on the open circuit voltage and heat dissipation of $Li/SOCl_2$ cells was studied. Lithium digestion was used to analyze the effect of heat treatment on cell anodes. Further, cell cathodes were characterized by BET single point nitrogen adsorption and mercury intrusion porosimetry prior to and following heat treatment.

INTRODUCTION

Lithium/thionyl chloride cells are widely used due to their high energy density and stable voltage. The high volumetric energy density of the oxyhalide system makes the lithium/thionyl chloride cells attractive for use in implantable medical devices, where the size of the power source is a key consideration. Prior to implant, medical devices require sterilization and, frequently, autoclave treatment is used rather than exposure to ethylene oxide. The use of autoclave sterilization demands that the power source and other device components be capable of withstanding exposure to elevated temperature without dramatic loss in function. This paper reports the results of an investigation of the effects of high temperature exposure (130°C) of lithium/thionyl chloride cells and cell components on cell performance.

EXPERIMENTAL

Cell Assembly

Case-positive experimental cells used in this study were housed in stainless steel, hermetically sealed prismatic cases having dimensions 45 x 23 x 8.6 mm. The lithium anode, constructed by pressing scraped lithium metal onto nickel screens equipped with leads, had a total surface area of 15.32 cm². Anodes were wrapped in a glass fiber separator (Mead Paper, Specialty Paper Division) and a polymer membrane (Zitex, Norton). Cathodes were formed by pressing a mixture of carbon and Teflon™ binder onto nickel screens. The cathodes were placed against the flat case walls on each side of the central lithium anode. The cells were vacuum-filled with either 0.4 or 1.0M lithium tetrachloroaluminate (LAC) in thionyl chloride (TC). The lithium chloride used to prepare the electrolyte was dried before use, while the aluminum trichloride and thionyl chloride were used as received. The cells had theoretical capacities of 3.2 Ah.

Similar cells were constructed with a lithium reference electrode, so that the individual contributions of the anode and cathode to the overall cell voltage could be assessed. These cells

34

were lithium-limited in design, having theoretical capacities of 1.5 Ah.

Cells were also constructed in which the 0.4M LAC in TC was fortified with either S_2Cl_2, SCl_2, or Cl_2 at levels of 5%, 5%, and 3% by weight. These additives were used as received.

Heat Treatment--Autoclave Simulation

All heat treatments referred to in this study consisted of a specified number of one hour heat cycles at $130 \pm 5°C$. The items being treated were removed from the forced-air oven and allowed to cool to room temperature between heat cycles.

Cell Storage and Discharge Performance

Cells were stored either at 37°C or at room temperature for periods of up to 959 days or were discharged under various resistive loads (0.301, 1, 2, and 5 kohms) at 37°C. Cell voltage and 1kHz impedance were measured intermittently, using an in-house, semi-automatic measuring system interfaced with a PRIME computer. The heat dissipation of cells on open circuit was measured periodically in a Tronac calorimeter (Model 351RA).

Effect of Heat Treatment on Electrolyte

I. Cell Discharge Testing

Ten battery cases were vacuum-filled with 0.4M LAC in thionyl chloride electrolyte, hermetically sealed, and subjected to five heat cycles. Additional electrolyte was stored at room temperature to function as a control. Twenty-four prismatic experimental cells were assembled to the point of filling. The heat treated electrolyte was used to fill twelve cells, while the remainder were filled with the control electrolyte. The completed cells were discharged under loads of 0.301 and 2 kohms.

Effect of Heat Treatment on Cell Anodes

Anodes were prepared by pressing lithium strips onto nickel screens. The weight of lithium used was recorded to 0.0001 g. The anodes were sealed into cells. The cells were vacuum-filled with thionyl chloride electrolyte, and subjected to either five 130°C heat cycles or stored at room temperature. After treatment, the cells were cut open and the anodes were removed, rinsed with thionyl chloride to remove any residual salts, and vacuum dried at 70°C. The remaining lithium was quantified, by reaction with water and titration of the resulting solution with hydrochloric acid using bromocresol green as indicator.

Effect of Heat Treatment on Cathode Material

I. Cell Discharge Testing I

Cathode plates were inserted into empty battery cases, filled with 0.4M LAC in thionyl chloride electrolyte and hermetically sealed. Half of the cases were then subjected to five heat cycles, while the remainder were stored at room temperature. A third group of cathode plates was stored dry at room temperature to serve as controls. The cases were cut open and the cathode plates were extracted. The plates were rinsed with thionyl chloride to remove residual salts. Lithium/thionyl chloride cells were built as described in **Cell Assembly**, using cathodes from each of the three groups. The cells were then discharged under either 0.301 or 2 kohm loads.

II. Cell Discharge Testing II

Cathode plates were inserted into either empty battery cases, those filled with only 0.4M LAC in thionyl chloride, or those with 0.4M LAC in thionyl chloride containing a small piece of lithium metal wrapped in glass fiber separator. These cells were either subjected to one heat cycle or stored at room temperature until use. The cases were cut open, and the cathode plates were removed and dried in a vacuum oven at 140°C. Mix was scraped from the screens of two cathode plates of each of the six groups and fresh cathode mix was pressed onto the screens. Lithium/thionyl chloride cells, incorporating a reference lithium electrode, were built using either the pretreated cathodes or screens from the six groups. These cells were placed under a resistive load (either 1 or 2 kohm) for one week. After removal of the load and a two or three day open circuit stand, the cells were pulsed using four 10-second pulse trains at a current density of 0.653 mA/cm^2, with a fifteen second rest period between pulses. Initial voltage measurements were taken using a Keithley electrometer and all data obtained during cell pulsing was recorded on a three-pen strip chart recorder. The resistive load was replaced after each weekly cell pulsing.

III. Specific Surface Area and Porosimetry Analysis

Cathode material from plates treated as described in the preceding section was scraped off the screens and analyzed by BET single point nitrogen adsorption using a Quantasorb instrument. Similar samples were analyzed by mercury intrusion porosimetry analysis using ASTM method D 4284.

RESULTS AND DISCUSSION

Cell Discharge Performance

Lithium/thionyl chloride cells are used in implantable medical devices which may be sterilized by autoclave. The effect that this type of heat treatment has on the performance of Li/SOCl$_2$ cells is, therefore, of interest. Figure 1 shows the performance of typical heat treated and non-heat treated thionyl chloride cells when discharged under 2 kohm loads. Table I lists the typical discharge capacities delivered by heat treated and non-heat treated cells placed under 0.3, 1, 2, and 5 kohm loads (currents of 12, 3.6, 1.8, and 0.72 mA, respectively; current densities of 0.78, 0.23, 0.12, and 0.05 mA /cm^2) . Using the F test, statistical analysis of the data obtained for cells discharged under 0.301 kohm loads revealed no significant difference between the capacities delivered by heat treated and non-treated cells. However, heat treated cells discharged under 1, 2, and 5 kohm loads delivered significantly less capacity than similar non-heat treated cells. The cause of this decrease in capacity was investigated by studying the effect of heat treatment on each individual battery component.

Effect of Heat Treatment on Electrolyte

Heat treatment of lithium/thionyl chloride cells at 130°C could alter the chemical composition of the electrolyte, which in turn could affect the discharge performance of these cells. While the generally accepted discharge reaction of the lithium/thionyl chloride couple is that presented in Equation 1,

$$4\,Li + 2\,SOCl_2 \text{-----}> \; 4\,LiCl\,(s) + SO_2\,(g) \; + \; S\,(s) \qquad [1]$$

other side reactions may also occur, especially at elevated temperatures. The most probable may include the thermal decomposition of thionyl chloride (1) (eq. 2), as well as the dissociation of

36

sulfur dichloride (1) (eq. 3).

$$4 \, SOCl_2 \, \text{------>} \, S_2Cl_2 \, + \, 2 \, SO_2 \, + \, 3 \, Cl_2 \qquad\qquad [2]$$

$$2 \, S \, + \, 2 \, Cl_2 \, \text{<------>} \, 2 \, SCl_2 \, \text{<------>} \, S_2Cl_2 \, + \, Cl_2 \qquad\qquad [3]$$

The formation of a number of sulfur chloride intermediates has been studied by several researchers. Blomgren et al. (2) cited gas chromatographic evidence of S_2Cl_2 in the electrolyte of partially discharged cells that was not present in fresh electrolyte. Carter and co-workers (3) analyzed fresh and fully discharged cell electrolyte also by GC. Significant amounts of S_2Cl_2 and SO_2, as well as the presence of Cl_2, was observed in discharged electrolyte. Their analysis of fresh electrolyte yielded the presence of some SO_2 and S_2Cl_2.

Chua et al. analyzed electrolyte from fresh, partially discharged, and fully discharged cells by IR (4) and proposed that the elevated level of sulfur monochloride found in discharged cells was due to a reaction between sulfur and thionyl chloride. Bailey studied the composition of electrolyte in partial cells by GC and GC-MS techniques (5).

I. Cell Discharge Testing

Samples of electrolyte subjected to high temperature cycles in the absence of cathode mix or lithium were used to fill prismatic cells. The open circuit voltages and performance of these cells were compared with cells filled with control electrolyte as listed in Table II. The cells filled with heat treated electrolyte had higher initial open circuit voltages (3.783 ± 0.013 V) than cells filled with the control electrolyte (3.699 ± 0.022 V). This may be due to the increased concentrations of Cl_2/HCl and SCl_2 in the heated electrolyte, as the potential of the Li/Cl_2 and/or Li/SCl_2 couple would be higher than that of the $Li/SOCl_2$ couple. The performance of the two groups of cells, however, showed no significant difference using the F test. Cells discharged under 0.301 kohm loads delivered 1.5 Ah, while cells which were discharged under 2 kohm loads generally delivered 2.9 Ah to a 2 volt cut-off.

Cells with salt concentrations of 0.4 and 1.0M LAC were placed on extended open circuit storage to determine the effect of heat treatment on open circuit potential. The voltage profiles of these cells versus time are shown in Figure 2. Generally, cell voltage increased with storage time for each group. This trend has been well documented in the literature (5-7). However, the voltage profiles of the two groups differed. By day 959, all cells containing heat treated electrolyte, regardless of the LAC salt concentration used, had higher open circuit potentials than cells filled with control electrolyte. In addition, the voltage appeared to be dependent upon the salt concentration in the electrolyte, as cells filled with 1.0M LAC in TC had higher potentials than cells filled with 0.4M LAC in TC.

The open circuit potential and heat dissipation of cells built with electrolyte fortified with S_2Cl_2, SCl_2, or Cl_2 were monitored as functions of open circuit storage at 25°C. The results are presented in Figures 3 and 4. It can be seen that initially the OCV fell into three groups: Cl_2-fortified; SCl_2-fortified and controls; and S_2Cl_2-fortified. By day 50, it appeared that the OCVs of cells containing either control or sulfur monochloride-fortified electrolyte were nearly identical and the two groups of cells continue to display the same OCV profiles. At approximately day 50, the OCV of cells containing sulfur dichloride-fortified electrolyte had substantially decreased. These cells appeared to have reached and maintained a plateau in

37

potential from day 100 to 325. Cells with chlorine-fortified electrolyte continue to have the highest open circuit potentials at 3.95 V, with a plateau in voltage occurring at approximately day 50. It should be noted that the cells with the highest OC voltages also dissipate the largest amounts of heat.

Effect of Heat Treatment on Cell Anodes

Although the cell construction is not initially lithium limited, consumption of lithium during heat treatment could also affect cell discharge performance. The results of the lithium digestions carried out on cell anodes which were and were not exposed to heat treatment are summarized in Table III. While exposure to thionyl chloride does expend some elemental lithium due to the formation of the LiCl passivation layer, the additional exposure to heat treatment does not affect the lithium loss. This is significant in that the reduced performance of lithium/thionyl chloride cells is therefore not due to a loss in available lithium for reaction with the electrolyte because of side reactions occurring with electrolyte decomposition products.

Effect of Heat Treatment on Cathode Material

I. Cell Discharge Testing I

Table IV contains a comparison of the discharge performance of cells built with heat treated cathodes, cathodes stored in TC electrolyte at room temperature, and cathodes stored dry. Statistical analysis of the data using the F test indicated that there is no significant difference among the cells discharged under a 300 ohm load. However, the analysis did reveal a significant difference in discharge performance among the cells when placed under 2 kohm loads.

II. Cell Discharge Testing II

Cells were built using cathode plates which had been subjected to various treatments of electrolyte and/or heat. From the results presented in Table V, it can be seen that the capacity delivered and capacity per gram of carbon by the six different groups of cells basically fall into two populations. The cells containing cathodes which had been previously exposed to thionyl chloride electrolyte delivered significantly less capacity than cells built with cathodes that had not been in contact with the electrolyte. Similar trends can be seen in Figure 5 in which RDC is plotted as a function of discharge capacity. The RDC is taken as the difference between the open circuit potential of the cell recorded just prior to pulsing and the fourth pulse minimum voltage divided by the pulsing current (eq. 5).

$$RDC = (OCV-P4min) / I \qquad [5]$$

The cells built with control and only heat-treated cathodes show very little change in their overall potential drops when periodically pulsed. The changes in overall potential and RDC are much greater, however, for cells constructed with cathodes pretreated in the presence of TC electrolyte. The data do not indicate a consistent trend in the effect of heat treatment on the performance of these cells.

Cells built in a lithium-limited design which incorporated a reference electrode were also intermittently pulsed. The results indicated that for the first pulse regime, the cathodic contribution to the overall voltage drop was greater than that recorded for the anode of cells built with cathodes either heat treated in the presence or absence of electrolyte. The anodic contribution was only slightly higher than the cathodic voltage drop in cells constructed with the control cathodes. In order to obtain a more complete profile of the electrode contributions to voltage drop, non-lithium limited cells were built with a reference electrode. In the non-lithium

38

limited design, all cells, regardless of pretreatment, exhibited larger anodic contributions to the overall voltage change up to 1.8 Ah capacity. After approximately 1.8 Ah of capacity was delivered, however, the cathodic contribution to the voltage drop became larger than the anodic contribution only for cells built with cathodes preexposed to thionyl chloride and this behavior continued through cell life. For cells built with the control and only heat-treated cathodes, the anodic contribution was always larger than that of the cathode throughout cell life, indicating that these cathodes were not affected by the treatment in a manner which would compromise cell performance.

Upon pulsing, all cells built with the screens from pretreated cathodes onto which fresh carbon mix was pressed exhibited anodic contributions to the overall cell voltage drop that were larger than that of the cathode up to a delivered capacity of 1.2 Ah. However, the overall voltage drop and those of the individual electrodes of these cells were approximately one-half to one-third of the magnitude of those recorded for cells in which entire pretreated cathodes were used. Up to 1.2 Ah of capacity, the cell profiles were identical to those of cells built using control and only heat-treated cathodes. After 1.2 Ah of delivered capacity, however, the cells exhibited pulse behavior reflective of the cathode treatment used. Cells built using screens from cathodes which had been preexposed to thionyl chloride electrolyte had increased cathodic contributions to the overall cell voltage drop, again noting that the voltage drop was only about one third of that of cells with pretreated cathodes. The anodic contribution remained larger than the cathodic for cells built with screens from control cathodes and those which had only been exposed to one heat cycle. While the data indicate that reduced performance in lithium/thionyl chloride cells is related to cathodic functioning, the results obtained for cells built with preexposed screens seems to indicate that reduced cell performance may also be due, in part, to a screen corrosion type of effect which may increase the RDC of the cell.

III. Specific Surface Area and Porosimetry Measurements

Changes in the physical properties of the cathode material can affect cell discharge performance. The possibility that autoclave procedures may alter the cathode pore structure was studied via single-point BET specific surface area and mercury intrusion porosimetry analysis. Table VI summarizes the results obtained for control and pretreated cathode material. The data indicate the existence of two populations-- material not exposed to thionyl chloride and that which has been preexposed to the electrolyte. Pressed or unpressed material not exposed to 0.4M LAC in TC had a specific surface area almost twice as large as that of material which had been exposed to TC. The porosimetry data also indicate that exposure to thionyl chloride changes the pore size distribution of the carbon mix. Material which has been exposed to electrolyte displays a significantly smaller intrusion volume (average of 37% less total volume) than material not exposed to TC electrolyte. Exposure of material to one heat cycle does not cause further reduction in the intrusion volume, nor in the median pore diameter.

CONCLUSIONS

The exposure of $Li/SOCl_2$ cells to heat treatment reduces cell performance and increases the internal impedance of the cell. Storage at 37°C leads to an increase in cell voltage, regardless of cell treatment, with the increase in potential being larger for the heat-treated cells. Electrolyte salt concentration also influences cell voltage. Use of 1.0M LAC leads to a higher voltage for heat treated cells, whereas cells containing 0.4M LAC evidenced a large increase in potential for the heat treated cells, but had a net lower voltage. Heat treated cells may also contain higher levels of sulfur monochloride than similar non-heat treated cells. Therefore, the higher cell voltages measured for cells containing 0.4M LAC in TC may be related to the sulfur monochloride content that increases with storage time. This species should preferentially discharge due to the higher voltage of the Li/S_2Cl_2 couple.

While storage of lithium in thionyl chloride does reduce the amount of elemental lithium present due to the formation of LiCl as a passivation layer, exposure of these partial cells to heat treatment does not further reduce the amount of lithium present. However, cells built with cathodes which were pretreated in the presence of catholyte did show decreased capacities on discharge and larger cathodic contributions to the overall cell voltage drop upon pulsing. This may be due to physical changes in the carbon material upon heat treatment in the presence of 0.4M LAC in TC as evidenced by a decreased specific surface area, intrusion pore volume, and median pore diameter, reducing the available active sites at which further reaction with the catholyte could occur during normal cell discharge. The reduced cell performance may also be affected, but to a much lesser degree, by corrosion of the cathode screen.

ACKNOWLEDGMENTS

The authors wish to acknowledge the contributions of D. R. Tuhovak, G. Beutel, K. Patterson, P. Sheehan, and D. Wesolowski to the work presented.

REFERENCES

1. R. C. Brasted, Comprehensive Inorganic Chemistry, Vol. 8, Pergamon Press, New York, NY (1973).

2. G. E. Blomgren, V. Z. Leger, T. Kalnoki-Kis, M. L. Kronenberg and R. J. Brodd, J. Power Sources, 7, 583 (1973).

3. B. J. Carter, R. Williams, A. Rodriguez, F. Tsay and H. Frank, Abstract 293, pp. 475 - 476, The Electrochemical Society Extended Abstracts, Fall Meeting, Detroit, Michigan, Oct. 17-21, 1982.

4. D. L. Chua, S. L. Deshpande and H. V. Venkatasetty, in "Battery Design Optimization," S. Gross, Editor, pp. 365-376, The Electrochemical Society Softbound Proceedings Series, Princeton, New Jersey (1979).

5. J. C. Bailey, in "Lithium Batteries," A. N. Dey, Editor, pp. 121-128, The Electrochemical Society Softbound Proceedings Series, Pennington, New Jersey (1987).

6. M. Babai, S. Bababigi and J. Bineth, in "Lithium Batteries," A. N. Dey, Editor, pp. 194-205, The Electrochemical Society Softbound Proceedings Series, Pennington, New Jersey (1984).

7. H. V. Venkatasetty and D. J. Saathoff, J. Electrochem. Soc., 128, 773 (1981).

Table I
Discharge performance of thionyl chloride cells
which have and have not undergone heat treatment.

Load (kohm)	Heat treatment	Discharge capacity delivered to 2.0 V (mAh)
0.301	Yes	1772
	No	1830
1.0	Yes	2323
	No	2845
2.0	Yes	2352
	No	2810
5.0	Yes	2405
	No	2700

Table II
Initial voltage and discharge capacities of thionyl chloride cells tested at 37°C.

Discharge load (kohm)	Electrolyte heat treatment	mAh to 2.0V	mAh to 1.7V	Initial voltage, V
0.301	Yes	1470	1500	3.771
0.301	Yes	1600	1650	3.787
0.301	Yes	1640	1680	3.804
average		**1570**	**1610**	**3.787**
0.301	No	1660	1680	3.706
0.301	No	1560	1590	3.718
0.301	No	1450	1480	3.676
average		**1557**	**1583**	**3.700**
2.0	Yes	2920	2930	3.771
2.0	Yes	2930	2940	3.779
2.0	Yes	2940	2950	3.788
average		**2930**	**2940**	**3.779**
2.0	No	2930	2940	3.725
2.0	No	2810	2815	3.672
2.0	No	2880	2900	3.694
average		**2873**	**2885**	**3.697**

Table III
Percentage of lithium depleted from partial cells during heat treatment or
storage at room temperature.

Heat treatment	% lithium utilized (± std. dev.)
Yes	8.3% ± 1.80
No	8.3% ± 1.70

Table IV
Average discharge capacities of thionyl chloride cells built with heat treated
and non-heat treated cathodes.
(n = 6 cells for each group)

Cathode treatment	Capacity to 2.0V (mAh)	
	0.301 kohm load	2.0 kohm load
Stored dry at room temperature	1453	2867
Stored in electrolyte at room temperature	1303	2223
Heat treated in electrolyte	1230	2623

Table V
Average cell performance as functions of cathode pretreatment.
(n = 3 cells for each group)

Cathode pretreatment	Code	Capacity delivered to 2.0V (Ah)	Ah/g carbon
Control - no TC, no heat	A	2.600	5.89
Control - heat only	B	2.630	5.94
0.4M LAC in TC	F	2.520	5.68
0.4M LAC in TC & heat	C	2.325	5.22
0.4M LAC in TC, Li	G	2.385	5.34
0.4M LAC in TC, Li, & heat	D	2.455	5.53

Table VI
Pretreated cathode material analysis.

Pretreatment code*	Specific surface area (m^2g^{-1})	Total intrusion volume ($cc\ g^{-1}$)	Median pore diameter (μm)**
A	53.999	4.2327	1.661
B	51.690	4.1102	1.387
C	27.940	2.6021	0.599
F	34.838	2.6582	0.460

*see Table V for complete description
**based on volume

Figure 1
Autoclaved and non-autoclaved cells discharged under 2 kohm loads.

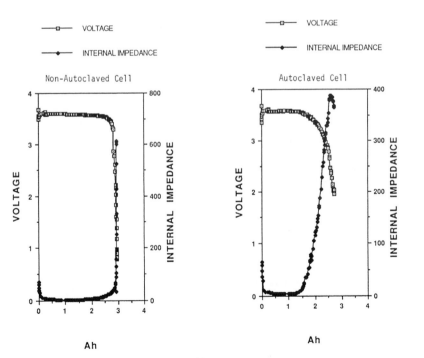

Figure 2. Average OCV at 37°C for heat-treated (HT) and non-heat treated (NHT) cells with different electrolyte LAC concentrations.

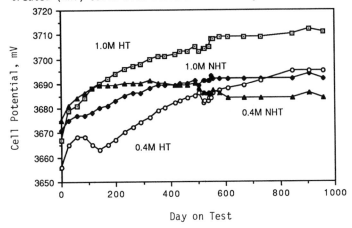

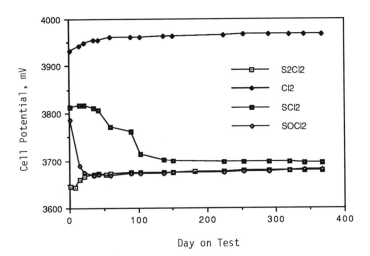

Figure 3. Average OCV as a function of storage at room temperature for non-heat treated cells with different electrolyte additives.

44

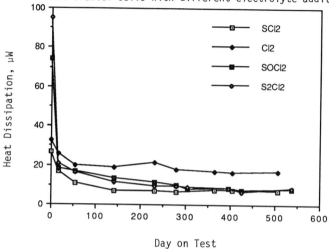

Figure 4. Average heat dissipation, recorded at room temperature, of non-heat treated cells with different electrolyte additives.

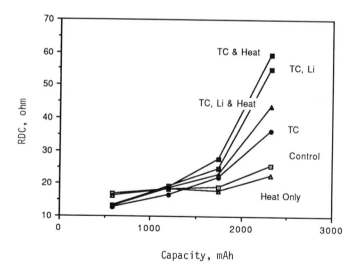

Figure 5. RDC as a function of capacity for cells built with pretreated cathodes.

FEASIBILITY OF A SUPER HIGH ENERGY DENSITY BATTERY OF THE Li/BrF$_3$ ELECTROCHEMICAL SYSTEM

Michael F. Pyszczek, Christine A. Frysz, Steven J. Ebel

Wilson Greatbatch Ltd.
10,000 Wehrle Drive
Clarence, NY 14031

ABSTRACT

The practical aspects of developing a lithium/bromine trifluoride battery have been studied. Efforts toward identifying materials for internal cell components via electrochemical testing techniques have resulted in a list of materials suitable for this application.

Prototype cells utilizing a spirally wound electrode configuration have been constructed and discharged. Through the use of currently available technology, however, the lithium/bromine trifluoride couple has not delivered energy density comparable to other high energy density systems.

INTRODUCTION

Bromine trifluoride (BrF$_3$) used in conjunction with lithium offers the possibility of batteries having very high cell voltages and energy densities. The liquid range of BrF$_3$ (8.8°C to 125.8°C) is well suited for application in ambient temperature lithium cells, and the self-ionization of BrF$_3$ to BrF$_2{}^+$ and BrF$_4{}^-$ suggests that the addition of electrolyte salts may be unnecessary. Consequently, the feasibility of developing a lithium bromine trifluoride battery has been under investigation [1,2].

The concept of a "Super High Energy Density Battery" of the Li/BrF$_3$ couple utilizing an antimony pentafluoride (SbF$_5$) electrolyte had been patented previously by the U.S. Navy [3]. More recently, Miles and coworkers at the Naval Weapons Center (NWC - China Lake, CA) have been reporting on experiments directed towards determining the suitability of BrF$_3$ as a lithium cell catholyte [4,5]. Miles' results suggest that it may be feasible to develop practical lithium cells of the Li/ BrF$_3$ couple that will have operating voltages above 4.5 V at current densities of 20 mA/cm^2.

To compliment the current work being done at NWC, Wilson Greatbatch Ltd. (WGL) proposed to perform feasibility studies on the practical aspects of lithium cell design utilizing BrF$_3$ depolarizer.

Studies focused on the following:

  (1) metals and alloys for battery containment;
  (2) cathode substrate/metal current collector couples;
  (3) hermetic electrical feedthroughs;
  (4) interelectrode separator material;
  (5) internal insulators;
  (6) cathode substrate materials;
  (7) long term lithium anode stability in lithium bromine trifluoride containing either lithium tetrafluoroborate or lithium hexafluoroarsenate salts; and,
  (8) energy density of prototype lithium bromine trifluoride batteries.

This report summarizes the test findings related to materials compatibility and performance of prototype Li/BrF_3 cells.

EXPERIMENTAL

The corrosion resistance of individual selected metals and alloys in BrF_3 was evaluated by means of immersion storage tests and by potentiodynamic polarization scans. All studies employed as-received BrF_3 obtained from Atochem North America.

BrF_3 Immersion Testing of Selected Metals and Alloys:

Initial compatibility testing involved storage of metal/alloy coupons immersed in BrF_3. Among the materials evaluated were Superferrit®, Shomac®, and 29-4-2, members of a class of high-chromium ferritic stainless steels. Thirty coupons cut from ten selected metals/alloys were cleaned, weighed, and placed in individual conical Teflon® PFA vials. Working in a glove bag containing a Teflon®-lined stainless steel tray, BrF_3 was added to each vial and the vials were capped. Approximately six cycles of evacuation and dry argon purge were carried out prior to filling and capping to minimize atmospheric contamination. Storage of the three groups of ten vials was for 2, 30 and 90 days. Upon removal, each coupon was washed in trichlorotrifluoroethane, reweighed, then examined via scanning electron microscopy (SEM). Surface condition was compared to virgin material.

The 90 day findings, shown in Table I, indicate that insignificant post-storage weight changes occurred. Except for 304 L stainless steel and mild steel which exhibited small areas of pitting corrosion, sample condition was comparable to that observed after 30 days of immersion.

TABLE I - Metal Evaluation Results

Sample	Pre/post test Wt.(g) 2 days	Pre/post test Wt.(g) 30 days	Pre/post test Wt.(g) 90 days
Mild Steel	.203/.204	.304/.302	.465/.466
304 L SS	.039/.040	.013/.012	.026/.026
316 L SS	.019/.020	.031/.030	.033/.034
Nickel 200	.293/.293	.408/.407	.361/.361
Superferrit®	.087/.087	.133/.133	.158/.158
Shomac®	.045/.045	.055/.054	.041/.041
Aluminum 1145-0	.031/.031	.024/.023	.020/.020
Hastelloy® G30	.507/.506	.657/.656	.605/.605
29-4-2	.072/.072	.104/.103	.099/.099
Monel 400	.473/.473	.192/.192	.294/.294

Potentiodynamic Polarization - Selected Alloys Versus Platinum Reference

Potentiodynamic polarization was employed to provide a qualitative means of comparing the behavior of ten selected metals in BrF₃. The potentiodynamic polarization procedure as outlined in the ASTM method G5-87 entitled "Standard Reference Method for Making Potentiostatic and Potentiodynamic Polarization Measurements" was followed with two exceptions: (1) the test cell and (2) the scan rate. An EG&G PARC model 273 potentiostat/galvanostat was used for the measurements and the data was analyzed via model 332 Softcorr corrosion Software.

The cell specified in ASTM G5-87 has a working capacity of one liter. After each run the solution is discarded. Because of the reactive nature of lithium bromine trifluoride, an alternate Teflon® cell with a 5 ml capacity was designed. The reference and auxiliary electrodes are held in place by a Gortex® gasket and an electrode holder. The test cell design allowed for blanketing the internal cell environment with argon at positive pressure to eliminate obscuring the test results by reactivity with air. The working electrode, a 0.625" diameter disk, is inserted against a Kalrez® O-ring. A hollow teflon screw with a nickel wire provides electrical connection to the system when assembled. The chamber is sealed with a Teflon® stopper.

To evaluate the performance of the cell, a potentiodynamic polarization scan of 430 stainless steel (SS) in 1N H_2SO_4 solution was completed in accordance with ASTM G5-87. The scan, however, was performed at 0.2 mV/s rather than the specified 0.167 mV/s. Since the results obtained were within the limits of the ASTM plot, the scan rate remained at 0.2 mV/s. Samples were cathodically then anodically polarized from -1000 mV to 2000 mV using platinum wire as reference and auxiliary electrodes. Current densities were measured over this potential range. Polarization characteristics were obtained by plotting the current response as a function of the applied potential via a log current function versus a potential semi-log chart.

Table II lists the measured corrosion potential (E_{corr}), corrosion current density (I_{corr}) and the calculated corrosion rate for the metals/alloys evaluated using platinum as the reference electrode. Response comparisons suggest Aluminum 1145-0 most resistant to bromine trifluoride with Nickel 200 and Shomac® as alternates.

48

TABLE II - Potentiodynamic Polarization Results - Platinum Reference Electrode

Material	E_{corr} (mV)	I_{corr} ($\mu A/cm^2$)	Corrosion Rate (MPY)
Superferrit®	- 452	.26	.08
Nickel 200	- 68	.07	.02
Shomac®	- 161	.07	.02
316 L SS	- 343	1.46	.39
Monel 400	- 500	1.46	.35
304 L SS	- 287	.90	.24
29-4-2	- 170	1.57	.43
Aluminum	-1021	.08	.01
Hastelloy® C22	- 608	1.27	.32
Lithium	-5143	429.30	725.21

Potentiodynamic Polarization - Metal/Alloy samples Versus Lithium Reference

   To facilitate comparison of candidate current collector materials, the potentiodynamic measurements of Table II were recalculated with reference to lithium. The lithium corrosion potential versus platinum was determined to be -5143 mV. Relating this value to the metal/alloy corrosion potentials versus platinum (Table II) allowed the desired recalculation yielding the E_{corr} value given in Table III. The current densities (I_{corr}) and corrosion rates of Table III were calculated on the basis of 5000 mV difference versus lithium since the open circuit voltage of carbon cathode substrate versus lithium in BrF_3 was found to average 4900 mV. These data suggest that Nickel 200 may be the best current collector candidate with Shomac®, aluminum and 29-4-2 as alternates.

TABLE III - Potentiodynamic Polarization Results - Lithium Reference Electrode

Sample	E_{corr} (mV)	I_{corr} ($\mu A/cm^2$)	Corrosion Rate (MPY)
Superferrit®	4691	.462	.14
Nickel 200	5075	- .291	in cathodic range
Shomac®	4982	.355	.09
316 L SS	4800	1.479	.40
Monel 400	4643	.828	.20
304 L SS	4856	1.034	.28
29-4-2	4973	.568	.16
Aluminum	4122	.316	.14
Hastelloy® C22	4535	.815	.20

Hermetic Electrical Feedthroughs

   Glass and ceramic-to-metal seals offer a means for hermetically sealing lithium batteries. Given that battery performance relies on retention of cell hermeticity, and that retention of cell hermeticity depends on preservation of seal integrity, storage of various sealing glasses/ceramics in PFA conical vials was performed. Results from weight loss analysis and visual observations are summarized in Table IV.

49

TABLE IV - GLASS AND CERAMICS IMMERSION RESULTS

SAMPLE	PRETEST WT. (g)	POST TEST WT. (g)	OBSERVATIONS
Fusite 435	0.526	0.529	etched
K Brown	0.104	0.07	partial disintegration
Fusite A 485	0.359	0.356	etched
as rec'd Cabal 12	0.063	0.064	slight discoloration
etched Cabal 12	0.064	0.065	slight discoloration
TM9 B	0.04		disintegrated
9013 Mansol	0.245	0.23	etched/discolored
9013 Corning	0.024	0.031	etched/discolored
Alumina	0.233	0.246	no discoloration
Fusite R1	0.243	0.253	etched/discolored
MSG 12	0.141	0.18	etched/discolored
TA-23	0.007	0.007	no discoloration
T clear	0.514	0.512	etched/discolored

Silicate glasses were observed to exhibit immediate bubble evolution when BrF_3 was introduced. After one month of room temperature storage, these glasses were visibly etched, with some samples disintegrating on handling. SEM examination revealed significant stress corrosion cracking. Of the glasses tested, Cabal 12, a nonsilica glass, and alumina, a ceramic, proved most promising.

Interelectrode Separator Materials

Successful separation of electrode components is crucial to both the performance and safety aspects of lithium cells. Favorable battery separator characteristics include: (1) resistance to degradation in the cell environment, (2) sufficient strength to facilitate cell fabrication, (3) porosity such that electrode separation is maintained while ionic transfer within the electrolyte is unimpeded, (4) surface energy such that electrolyte wettability and absorption are augmented, (5) sufficient thickness to maintain interelectrode separation without impedance to battery high performance, and (6) freedom from contaminants and reactive species capable of adversely affecting cell performance.

The following separators were evaluated: (1) Whatman BSF45, (2) Kaowool 90105-01, (3) Hollingsworth and Voss (H&V) BG03013 LN 96835, and (4) Rayperm 200/60. Whatman and H&V are glass fiber separators. Whatman is binderless and H&V contains an acrylic binder. Kaowool is primarily a ceramic fiber separator. A small percentage of glass fiber is also present. A latex binder is used for fabrication. The Rayperm® separator, supplied by Raychem Ltd., is fabricated from Tefzel®, ethylenetetrafluoroethylene (ETFE). It is heat sealable, has low surface energy and is not easily wetted by water. Compatibility was studied by adding 0.10 ml of bromine trifluoride to a 0.250" by 1.000" sample. The Whatman separator disintegrated within 2 seconds. A pyrophoric reaction was observed for the Kaowool. The Hollingsworth and Voss separator reacted with incandescence. No visible reaction was observed for the Rayperm separator, consequently, this material was placed in a teflon PFA vial into which bromine trifluoride was added. Storage was as previously described for the metal samples. The Rayperm separator was visually examined through the vial after one month of storage. The electrolyte remained clear and the sample appeared pliable and in good condition. The sample was returned to the PFA vial for 180 day storage.

50

Given the favorable results obtained with Rayperm®, additional compatibility testing was initiated. In addition to the original storage sample, two groups of twelve samples were prepared from stock 1.75 inch wide by 0.002 inch thick material. Group 1 comprised samples cut 1" wide by 7" long in the material machine direction (MD) while group 2 samples, 1" X 1.75", were cut in the material transverse direction (TD). Three samples from each group were stored at room temperature for 1 day, 2 weeks and 1 month immersed in BrF_3 with three virgin samples serving as control. A capped 2" high X 1" diameter PFA vial served as the storage vessel. Upon removal from the storage vial, all were found to be fully intact with no evidence of embrittlement.

To evaluate sample degradation, material tensile properties before and after BrF_3 storage were determined using an Instron Model 1130 Universal Tensile Machine. Ultimate tensile strength (UTS) and elongation to failure (%E) were determined at a crosshead speed of 1.0 inch per minute. Gage length was 5" in the MD and 1" in the TD. Degradation of the test samples was not apparent. Table V summarizes the tensile test results.

TABLE V - Tensile Test Results for Room Temperature Storage of Tefzel® in BrF_3

SAMPLE	TD/UTS	MD/UTS	TD/%E	MD/%E
Virgin 1	1.21	7.04	21.9	359.4
Virgin 2	1.32	6.60	28.1	281.3
Virgin 3	1.21	8.36	25.0	421.9
1 Day 1	1.21	7.48	27.5	406.3
1 Day 2	1.10	6.38	25.0	312.5
1 Day 3	1.10	6.60	25.0	359.4
2 Wks 1	1.21	7.04	21.9	359.4
2 Wks 2	1.21	7.48	21.9	390.6
2 Wks 3	1.10	6.16	21.9	312.5
1 Mo 1	1.21	6.16	28.1	281.3
1 Mo 2	1.21	4.84	25.0	250.0
1 Mo 3	1.21	7.04	21.9	343.8

Although the sample size was small, the data obtained suggests that, at room temperature, degradation of Tefzel® in BrF_3 is not imminent; this material may be viable for use as separator for a Li/BrF_3 cell.

Internal Insulator Materials

In addition to the interelectrode separator, practical battery design incorporates other internal insulator components. Bromine trifluoride is known to react violently with many organic compounds. The literature suggests the following fluoroorganics compatible with BrF_3: polytetrafluoroethylene (PTFE), polyfluoroethylenepropene (FEP), and polyperfluoroalkoxyethylene (PFA). These along with polyethylenetetrafluoroethylene (Tefzel®) and polyethylenechlorotrifluoroethylene (Halar®) were evaluated for long term stability in BrF_3 by immersion storage tests. Ten coupons were cut, cleaned, sized, weighed, and placed in individual conical Teflon® PFA vials. The vials were capped and stored for 90 days. Upon removal, each coupon was washed in trichlorofluoroethane, reweighed, then examined via SEM. Sample condition was compared to virgin material. All coupon weights, sizes and configurations remained unchanged when compared to pre-storage conditions. SEM examination revealed no evidence of crazing or stress corrosion cracking. Elastomeric properties appeared unaffected by storage.

Long Term Lithium Anode Stability

Because lithium is strongly electropositive, leading to high voltages when coupled with suitable cathode materials, and has high gravimetric and volumetric specific capacity, it is a highly desirable anode material. Lithium, however, forms a highly protective passive film in BrF_3, as it must in any cell having useful shelf life. Such film may be composed of a dense, very thin inner layer topped by a thicker, more porous outer layer. If the passive layer formed in a given electrolyte becomes excessively protective, severe voltage delays may be observed.

For the initial assessment of lithium stability in BrF_3, a 1.000" by 0.250" coupon, 0.011" thick was stored in BrF_3 for 90 days. Immediately upon immersion, a gray film formed on the lithium. After 90 days storage, the intensity of the gray film remained unchanged. Film formation appeared uniform with an apparent thickness of approximately 0.001". To determine the effect of film formation on cell performance, additional study is required.

Prototype Battery Energy Density

Test Cell Construction. A spirally wound 1/2 AA configuration was chosen as the test vehicle for initial prototype cell development efforts. The cathodes consisted of a high surface area carbon cathode formed by compression of a dry carbon black/PTFE mixture onto an expanded Superferrit® grid. A tab of similar material was resistance spotwelded to the current collector grid to act as the cathode lead. The electrochemical surface area of this design was determined to be 14.5 cm^2.

The anode was comprised of a section of pure lithium metal into which a tab of Superferrit® stock was pressed. The single piece of tab stock served as the current collector and external lead. Electrical insulation of the electrodes was accomplished through the use of a two layer separator arrangement. A microporous Tefzel® (ETFE) film was located adjacent to the carbon cathode to provide a positive barrier to dislodged carbon particles. The second layer was made up of a woven Tefzel® cloth which imparted a high degree of mechanical strength to the separator assembly and was located facing the anode. In addition to being tear resistant the compatible surface energy of the ETFE fibers allowed the bromine trifluoride catholyte to be readily absorbed and transported into the inter-electrode gap.

After the spirally wound electrode stack was inserted into the cell case, a two layer disk of the woven Tefzel® was positioned above the stack. This disk served as a diffuser plate to prevent the electrolyte stream from dislodging carbon during the filling operation. The cell case consisted of a Teflon® (PFA) round bottomed test tube 2.50" long with an outside diameter of 0.625" and a wall thickness of 0.030". After insertion of the internal components, the Superferrit® electrode leads were directed against the internal cell wall and fixed in place by insertion of the tapered header assembly. The header, which provided the gas tight seal for the cell, consisted of a tapered Teflon® (PTFE) stopper into which had been inserted two 0.125" O.D. PFA tubes used for introduction of catholyte in to the cell.

Equipment used for filling the cells is diagramed in Figure 1. Following assembly and closure of the test cell in a dry atmosphere it was placed behind a Lexan® safety shield and the pressure transfer tube was connected to the 3-way valve. The filling/vent tube was inserted into the electrolyte reservoir to which is added the electrolyte solution. The 3-way valve is opened to allow a flow of high purity argon to pass through the cell and through the solution thus removing any volatile impurities caused by hydrolysis reactions. After the electrical connections had been made between the measuring equipment and the electrode leads, the valve is rotated to connect the pre-evacuated low pressure reservoir with the test cell. The contents of the electrolyte reservoir are then drawn into the test cell. The valve is finally positioned to allow a slow stream of argon to sweep through the cell. This arrangement prevents a pressure build-up of reaction off-gasses to occur within the cell while maintaining an inert internal environment. Secondary safety shielding is than put in place and the cells electrochemical performance is evaluated.

The BrF_3 depolarizer solutions employed were straight BrF_3 (Atochem North America) used as received, or BrF_3 mixed (stirred for at least one hour and the solution decanted from the insolubles) with either $LiBF_4$ (Morita - Japan) or $LiAsF_6$ (Lithco Lectro-Salt). These salts were dried overnight in vacuum at 120°C before use. The solutions were prepared by stirring approximately 0.3 g of each salt with 10 to 15 g of BrF_3 under an argon stream. Slight effervescence accompanied this process for both salts and was observed to taper off, but not cease, over about two hours time. The effervescence was more pronounced with the $LiBF_4$ solutions than with the $LiAsF_6$ solutions. Although the actual solubility of these salts was never determined, in all cases it was estimated that a solution of less than 1% by weight of the solute was formed. A possible explanation for the effervescence could be the evolution of BF_3 or AsF_5 gas to form $Li^+ + BrF_4^-$ in-situ with the reaction proceeding more rapidly in the case of $LiBF_4$. A literature search could produce no record of a material of this composition being isolated previously, and we offer this hypothesis as a possibility without confirmation at this time. The only evidence for any reaction or dissolution we have is that there was a positive effect on the test cell discharge results.

RESULTS AND DISCUSSION

Six test cells of the 1/2AA size were successfully discharged through constant resistive loads of either 20Ω (three - one each of straight BrF_3, $LiBF_4/BrF_3$, $LiAsF_6/BrF_3$) or 250Ω (three - same). Figures 2 and 3 display the discharge curves for these test cells. Under 20Ω load (Figure 2) the discharge voltage sloped down rapidly over a time-frame of up to one hour. All of the test cells fell below 3 volts within 2 minutes of the initiation of discharge with the straight BrF_3 test cell falling off the fastest. At 3 volts a 20 Ω load corresponds to a discharge current density of approximately 10 mA/cm^2. At this rate, the highest capacity achieved was with either of the lithium salts ($LiBF_4$ or $LiAsF_6$) being employed. Figure 2 also shows that the capacity to 2 V was approximately 13 to 15 mAh

53

with the salts, and approximately 4 mAh without.

Under the less stressful 250 Ω load (Figure 3) the cells containing a lithium salt additive maintained a voltage above 4 V from 80 minutes to 160 minutes. At 4.5 V a 250 Ω load corresponds to a current density of 1.2 mA/cm². Without the salt the cell could only manage a voltage slightly above 2 V (this test was repeated with similar results). The capacity achieved to a 4 V cutoff was 45 - 50 mAh with the $LiBF_4$ salt and approximately 25 mAh with $LiAsF_6$ additive. To a 2 V cutoff the corresponding capacities were 65 and 30 mAh respectively. This would suggest a surprisingly low energy density estimate of only 0.065 Wh/cm³ for 1/2 AA size cells of the Li/BrF_3 couple with lithium electrolyte salt added - over an order of magnitude lower than is commonly realized in Li/oxyhalide cells of similar size.

Melvin H. Miles and coworkers of the NWC, have reported test cell results [4,5] using laboratory test cells which utilized 1-2 cm² Li disk anodes immersed in straight BrF_3 (and also BrF_3 containing NaF electrolyte salt), a carbon cathode disk, and a 3 cm diameter cylindrical platinum screen as reference electrode. These test cells were discharged using constant current at rates corresponding to from 5 to 50 mA/cm². At 5 mA/cm² they achieved a normalized capacity of approximately 18 mAh/cm². Our cells (with $LiBF_4$ additive) discharged at 1.2 mA/cm² and normalized in this way would achieve a value of only 3 mAh/cm². There were many differences in the details of the NWC test cells and the WGL test cells. It is not known what part these differences played in the discrepancies, although it is worth mentioning that the NWC test cells were of the button style and utilized stainless steel springs for electrode contact perhaps helping to keep a tight interfacial separation between anode and cathode. The WGL cells were of the cylindrical wound element type, and the winds were relatively loose (compared with normal commercial product) for safety reasons. The NWC group, as well as the investigators at EIC have also suggested that the main failure mode for the Li/BrF_3 cells is the severe passivation of the lithium surface while in contact with the BrF_3 depolarizer.

One final comment from a safety standpoint. It should come as no surprise given the reactive nature of the BrF_3 material that there were some anxious moments in the laboratory. Of ten test cell experiments, two ended abruptly in explosions. These explosions both occurred within seconds of filling the test cells. One of the explosions was proceeded by a notable flash near the top of the cell stack (visible due to the transparent PFA tubes used as cases). The other explosion occurred with no visible precursor. Both explosions occurred within 1 to 3 feet of the investigators. There were no injuries, nor extensive damage as a result of these incidents. The authors believe that the lack of injury or major damage was due to two main factors: (i) excellent shielding, and (ii) the nonmetal cases (tubes) being employed for cell containment. After the two incidents the experiments were changed slightly by incorporating a porous teflon diffuser plate at the top of the cell stack to prevent the rush of liquid BrF_3 during filling from breaking off chunks of the carbon cathode, and by fabricating the anodes just prior to the experiment (not a day or two ahead of time) because of a concern about lithium nitride (Li_3N) formation in the anode during the waiting period. It is not known for certain what effect these changes had on the safety of the test cells, but there were no additional safety incidents for the remaining seven consecutive test cell experiments.

CONCLUSIONS

The results obtained during this effort lead to the following conclusions:

1. It is feasible to conduct Li/BrF_3 primary cell research on spirally wound element cell configurations.

2. There are suitable materials available for Li/BrF_3 cell construction. Among these are...
a) cell separators - fluoropolymer films or woven fabrics, binderless nonsilica containing glass (or ceramic) fiber paper,
b) containers - fluoropolymers, nickel, aluminum, corrosion resistant ferritic stainless steels,
c) electrode current collectors - same metals as in "b,"
d) insulators - same as in "a,"
e) hermetic insulating feedthrough - nonsilica containing glasses or ceramics.

3. The use of suitable lithium salts such as $LiBF_4$ and $LiAsF_6$ improves the discharge characteristics of Li/BrF_3 cells.

4. Using current technology, the Li/BrF_3 couple has not delivered energy density comparable to other high energy density lithium systems.

5. Practical Li/BrF_3 cell development will require further extensive fundamental electrochemical cell research. This is mostly because of the extreme reactivity of the BrF_3 material and the need to maintain a safe experimental environment for cell fabrication and testing - this in excess of what is normally required of more common lithium cell research and development.

REFERENCES

1. Pyszczek, M.F., Ebel, S.J., Frysz, C.A., Paper #29 presented at the 180th Meeting of the Electrochemical Society, October 13-17, 1991, Phoenix, AZ

2. G. L. Holleck, G. S. Jones, ECS Battery Division Extended Abstracts of Fall Meeting, Seattle, WA, 1990.

3. Goodsen, F.R., Shipman, W.H., McCartney, J.F., U.S. patent number 4,107,401, 1978.

4. Park, K.H., Miles, M.H., Bliss, D.E., Stilwell, D., Hollins, R.A., Rhein, R.A., *J. Electrochem. Soc.*, 135, 2901 (1988).

5. Park, K.H., Stilwell, D.E., Bliss, D.E., Hollins, R.A., Miles, M.H., "A Comparison of Lithium, Magnesium and Calcium Anodes in Bromine Trifluoride," *Electrochem. Soc. Extended Abstracts,* 88-2, 30, 1988.

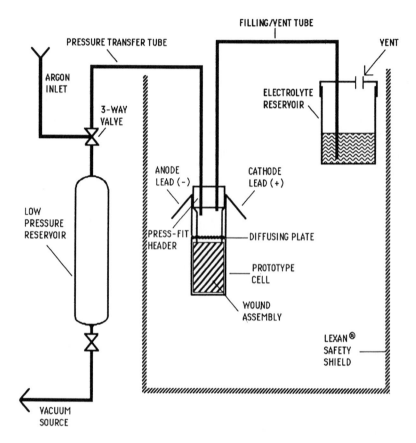

Fig. 1. Prototype Cell Filling Apparatus

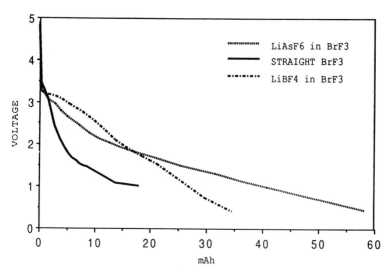

Fig. 2.   20Ω DISCHARGE of 1/2 AA CELLS

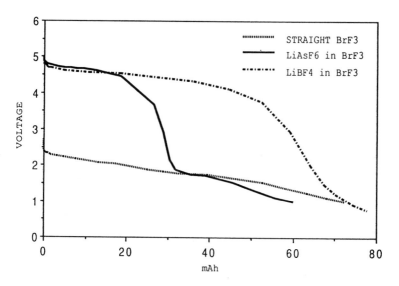

Fig. 3.   250Ω DISCHARGE of 1/2 AA CELLS

AN ENHANCED REDOX PROCESS OF DISULFIDE COMPOUNDS AND THEIR APPLICATION IN HIGH ENERGY STORAGE

Katsuhiko NAOI and Noboru OYAMA

Department of Applied Chemistry, Faculty of Technology
Tokyo University of Agriculture & Technology
Koganei, Tokyo 184, JAPAN

An enhanced redox process was found for one of the disulfide compounds, 2,5-Dimercapt-1,3,4 thiaziazole(DMcT), on an electropolymerized polyaniline(PAn) thin film. The behavior was investigated for its electrocatalytic characteristics by using cyclic voltammetry, AC impedance and quartz crystal microbalance techniques. The enhancement effect of DMcT by polyaniline could be plausibly explained by the possible formation of adducts on the nitrogen sites in polyaniline chain. Charge-discharge tests were performed on lithium cells with using polyaniline/carbon/DMcT composite cathode and gel electrolyte. The energy density for a tested Li/(DMcT+PAn) cell(>220 Wh/kg) far exceeded those of Ni/Cd or Ni/H$_2$ batteries, which is suggestive of the potential use in practical application as the rechargeable lithium battery system at ambient temperature.

INTRODUCTION

Very recently, a novel energy storage system using disulfide compounds as cathode material was proposed by Visco et al(1,2). A series of compounds having -SH groups in the organodisulfide compounds serves as the function of high energy storage battery system. Such an battery system have enormous energy density with respect to unit weight and/or volume per cathode active materials. In fact, disulfide compounds such as DMcT and TTcA have high theoretical energy density of more than ca. 1000 Wh/kg-cathode compared to the other further battery materials like conducting polymers or intercalation compounds(see Fig.1). The research of authors has been focused on series of compounds which have the following functional group in the moieties.

As shown in Fig.2, the compounds include DMcT, DTU, TTcA, and others. Although disulfide compounds inherently have a high capacity for energy storage, the kinetics for the redox reaction was found to be rather slow which is regarded as one of the serious drawbacks of these compounds. Among those, DMcT shows the most facile redox reaction and therefore is the most promising cathode material which can be utilized in rechargeable lithium cells as far as the authors are concerned.

In general, disulfide compounds shows redox reaction which seems to accompany structural change among monomers, dimers, and polymers. Thus, when the compounds are used as a cathode material in secondary battery, charging process induces polymerization(or dimerization, depending on the applied potential), and the discharging process causes depolymerization (or monomerization). Yet, the electrochemical properties

58

of these compounds were not much investigated in spite of the substantial interests by battery industry.

Application of organodisulfide compounds as an effective cathode material specifically in conjunction with solid polymer electrolyte has been proposed(1,2). For such an application of organodisulfide compounds, the redox process should be reversible for better performances in discharging or charging with efficient coulombic delivery. To enhance the redox process of the disulfide compounds, transition metal phthalocyanines, specifically cobolt phthalocyanine(CoPc) has already been found to serve as an electrocatalyst for the redox process(2). It was considered that the enhancement may be responsible for the CoPc adsorbed on the electrode. As another effective electrocatalyst for the redox process of disulfide compounds, the authors found electropolymerized polyaniline(3). The authors are proposing in the present article that an electropolymerized polyaniline catalyzes the redox of DMcT by forming adducts on the nitrogen atom in the polyaniline backbone.

EXPERIMENTAL

Disulfide compounds, 2,5-Dimercapt-1,3,4-thiadiazole(DMcT) were used after purification by crystallization. Acetonitrile was distilled at atmospheric pressure. TBAP was used as and supporting electrolyte without further purification. Cyclic voltammograms were obtained on a usual potentiostat HAB-151(Hokuto Denko Ltd.) at a scan rate of 50mV/s under an inert N_2 gas. DMcT(purchased from Aldrich Chemicals) was used after purification as described previously(5-7). Reagent grade tetrammonium perchlorate (TBAP)(Tokyo Kasei Co., Ltd.) was used as the electrolyte. Acetonitrile(AN) was used as a solvent after purification by distillation under reduced pressure.

A standard three-electrode, two compartment electrochemical cell was used for all the electrochemical experiments including cyclic voltammetry, pulse voltammetry, and electrochemical quartz crystal microbalance measurements. Potentials were referred with respect to Ag/Ag^+(0.01 M $AgNO_3$) electrode in the same solution. All experiments were run under a nitrogen atmosphere at room temperature(25.0 +1.0 ^oC).

RESULTS AND DISCUSSION

In acetonitrile, the electrochemical response of DMcT was investigated by cyclic voltammetry(see Fig.3). In Fig.3(C) voltammograms represent redox process corresponding to a dimer-monomer conversion at bare carbon electrode. Broadening between peak potentials of the redox couple indicates a slow reaction, which could be assigned as a quasi-reversible electrode process by our previous work(4). SEM micrographs presented in Fig.4 clearly demonstrate the evidence of the formation of polymeric products(at 0.5V vs. SSCE) from DMcT solution on BPG substrate. Setting the same electrode at -0.7V caused a complete depolymerization or detaching of coupled DMcT from the substrate and nothing was left on the surface. This can be cycled repeatedly, and the process of the reaction is quite reversible. In contrast, at polyaniline film coated electrode as shown in Fig.3(A), the cathodic peak at -0.65V shifts to a positive potential more than 450mV. By utilizing such a polymer-coated cathode, the large current density will be expected. Similar enhancement effect was observed at other electropolymerized conducting polymers which are based on aromatic amino compounds, for example, poly(o-phenylenediamine); the results are not presented here.

Similar experiment has been done on methyl-substituted polyaniline on N sites. Figure 5 shows typical cyclic voltammograms for DMcT on poly(N-methylaniline)(PMA) and poly(NN'-dimethylaniline) (PDMA). To ensure the formation of adducts or accumulation of DMcT, cyclic voltammetry was firstly done in supporting electrolyte solution without DMcT(A), then immersed in the solution containing DMcT(B), and finally immersed back in the first solution(A'). Neither of the voltammograms in (B) show any favorable potential shift which indicates kinetic enhancement, or the narrowing of the peak potential separation. The behavior could be explained by considering the formation of adducts between DMcT and nitrogen sites in the polyaniline moieties. Only at polyaniline-coated electrode, DMcT shows substantial shift to anodic direction. On the other hand, Anodic peak appeared to remain basically unchanged. The anodic shift of DMcT cyclic voltammogram is well consistent with the behavior shown by us in the previous papers concerning polyaniline.

As shown in Fig.6, impedance spectra for polyaniline in the presence or absence of DMcT also supports the adduct formation(6). The polyaniline film was successively immersed in the same solutions in the same procedure as in the cyclic voltammetric measurement. Polyaniline firstly showed larger semi-circle in the solution containing only supporting electrolyte(see Fig.6(A)). After the polyaniline film was immersed in the solution containing both supporting electrolyte and DMcT, the semi-circle became smaller(see Fig.6(B)), indicating that the electron transfer rate is enhanced in some way or another. Although the polyaniline film was immersed back to the first solution, the diameter of the semi-circle remain unchanged. This means that the polyaniline has changed completely from the native form by immersing in DMcT solution.

The schematic for the redox mechanism for DMcT on polyaniline is shown in Fig.7. The scheme involves a possible radical-radical coupling between N-sites in polyaniline and -SH sites in DMcT at higher oxidation stage maybe at the second oxidation potential for polyaniline. Once they make adducts, they behave like different kind of redox couples, which is consistent with the behavior in cyclic voltammograms and ac impedance spectra. Further studies on DMcT by spectroscopic analysis such as IR, Raman, and XPS are currently underway to elucidate a complete model for adduct formation.

QCM measurement on DMcT seems to support the adduct formation as well(see Fig. 8). The frequency shift vs. charges consumed during oxidation and/or reduction of DMcT/PAn redox system. PAn film has higher $\Delta f/Q$ slope compared to that of PAn/DMcT electrode system. This could be explained by assuming the schematic described above. For PAn film in non-aqueous media, only mobile species would be anions or dopantswithin the potential width tested here. On the other hand, once DMcT makes adducts with nitrogen sites in PAn chain, there becomes increased possible mobile species besides anions. The possible species to be mobile in this case are electrolyte anions(ClO_4^-), disulfide anions($^-SRS^-$) anions, electrolyte cations(Li^+) and others such as solvent molecules and DMcT monomers. Among the above species, the oxidation process accompanies anion doping which contributes to the mass increase, and cation undoping process associated with the crosslinking process which leads to the mass decrease. The later process which leads to the loss in mass of the film electrode can be considered to lower the $\Delta f/Q$ slope in Fig.8. The fact could be more or less explained by postulating the formation of adducts between DMcT and PAn films.

In Fig.9, are presented typical discharge curves for Li/SPE/(DMcT+PAn) cells with different cathode composition. The discharge curves were obtained at a rate of $0.1mA/cm^2$ after sufficiently charging at a constant voltage of 4.05 V. The dashed line shows higher discharge voltage compared to the other cathodes which do not contain PAn in the cathode composite. There is a clear indication that PAn works to promote the charge/discharge reaction of DMcT/PAn cathode. It should also be noted that the discharge curves generally have two plateau regions which can be assigned to polymer-dimer conversion and dimer-monomer conversion, according to our previous works(4-7).

Table 1 summarizes the cell performances including cathode utilization, energy density calculated per cathode active materials and the whole cell(7). The tested cell composed of lithium anode(80 µm thick; d=0.55 g/cm^3), SPE electrolyte(50 µm thick;

d=1.3 g/cm^3) and the composite cathode(170 μm thick, d=1.7 g/cm^3). The cell with (DMcT+PAn+SPE) cathode delivered more than 303 Wh/kg-cathode at relatively slow rate of 0.1 mA/cm^2. Even at higher rate of discharging(0.8 mA/cm^2), the energy density still amounts to ca. 76 Wh/kg-cathode. At a rate of 0.1mA/cm^2 of discharging, the speculated specific energy per cell would be 220 Wh/kg, which far exceeds those for the conventional cells such as Ni/Cd and Ni/H$_2$ batteries. Further studies on the cyclability or the self-dischargeability of Li/(DMcT+PAn) cells are currently underway.

Tab. 1 Calculated energy densities for Li/SPE/DMcT cells(7).

Cathode composition		DMcT+PAn+SPE	DMcT+Carbon+SPE
Cell impedance (Ω)		450	340
Capacity	0.1 mA/cm2	>94	96
	0.2	60	49
utilization (%)	0.8	25	17
Energy density	0.1 mA/cm2	>303	235
per cathode	0.2	183	122
(Wh/kg-cathode)	0.8	76	37
Speculated energy	0.1 mA/cm2	>220	172
density per cell	0.2	134	89
(Wh/kg-cell)	0.8	59	27

CONCLUSION

An enhanced redox process of one of the disulfide compounds, DMcT was observed at electropolymerized polyaniline electrode. According to the cyclic voltammetric and ac impedance measurements of the redox of DMcT, the enhancing effect can be attributed to the formation of adducts between -SH group and N sites in the polyaniline moieties. The energy density for a tested Li/(DMcT+PAn) cell(>200 Wh/kg) exceeded those of Ni/Cd or Ni/H$_2$ batteries by 4 to 5 times. Therefore it is revealed that the combination of DMcT and Polyaniline is of potential use in practical application as an ambient temperature rechargeable lithium battery material. Migration problem of DMcT either to current collector or to the counter electrode would also be expected to diminish by sticking DMcT in polyaniline chain as adducts.

ACKNOWLEDGEMENT

This work was supported in part by TEPCO Research Foundation and in part by The Japanese Ministry of Education. Travelling fee for the attendance to the meeting was funded by C & C research foundation. The authors acknowledges the technical support from T. Sotomura, H. Uemachi and K. Takeyama at Matsushita Electric Industrial Co. to estimate the cell performance.

REFERENCES

1.   M.Liu, S.J.Visco, and L.C.De Jonghe, *J.Electrochem.Soc.*,136, 2570 (1989).
2.   M.Liu, S.J.Visco, and L.C.De Jonghe, *ibid* , 137, 750 (750).
3.   T. Ohsaka, K. Chiba, N. Oyama, Nihon Kagaku Kaishi, 3, 457 (1986);
4.   K. Chiba, T. Ohsaka, Y. Ohnuki and N. Oyama, *J. Electroanal. Chem.*, 219, 117 (1987).
5.   K, Naoi, M. Menda, H. Ooike and N. Oyama, *Proc. 31st Bat. Symp. Jpn.*, p.31 (1990).
6.   K. Naoi, M. Menda, H. Ooike and N. Oyama, *J. Electroanal. Chem.*, in press.
7.   T. Sotomura, H. Uemachi, K. Takeyama, K. Naoi and N. Oyama, *Electrochim. Acta*, submitted (1991).

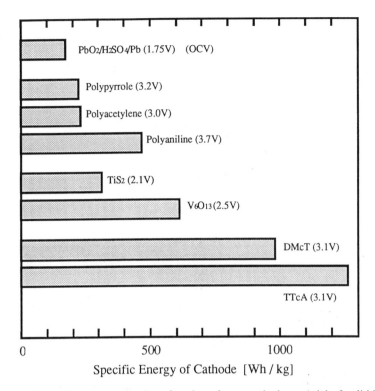

Fig.1   Theoretical energy density of various future cathode materials for lithium batteries. The materials include conducting polymers, intercalation compounds and two typical disulfide compounds(viz., DMcT and TTcA).

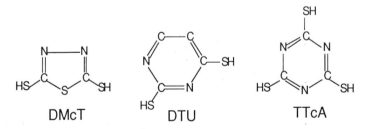

Fig.2   Typical disulfide compounds having the S-C-N group in the molecule.

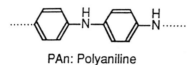

PAn: Polyaniline

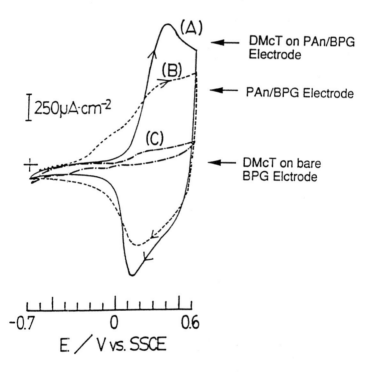

(A) ← DMcT on PAn/BPG Electrode

(B) ← PAn/BPG Electrode

250μA·cm⁻²

(C)

← DMcT on bare BPG Elctrode

E / V vs. SSCE

-0.7   0   0.6

Fig.3   Typical cyclic voltammograms for DMcT(5mM) on bare and polyaniline-coated electrode in acetonitrile solution containing 0.1 M LiClO₄.

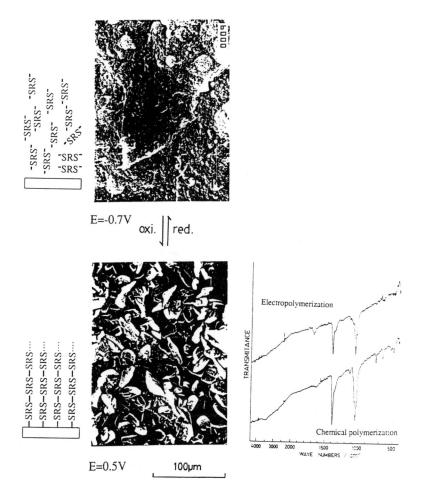

Fig.4   SEM micrographs of BPG surfaces polarized at 0.5V and 0.7V in the same
solution as in Fig.2.

65

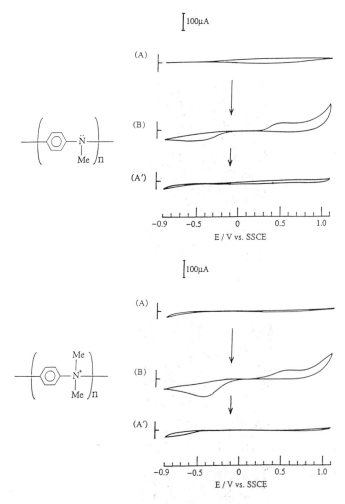

Fig.5   Cyclic voltammograms for DMcT on poly(N-methylaniline) and poly(NN'-di-
         methylaniline) coated BPG electrodes in the solutions in the presence(B) or
         absence(A ; A') of DMcT. Scan rate: 100 mV/s. Polyaniline films were prepared
         in 0.1 M respective monomer + 0.5 M Na_2SO_4(pH=1.03) by cycling potential
         between -0.8 and +1.2V for 5 cycles.
         (A) 0.1 M $LiClO_4$/AN
         (B) (5 mM DMcT + 0.1 M $LiClO_4$)/AN
         (A') 0.1 M $LiClO_4$/AN

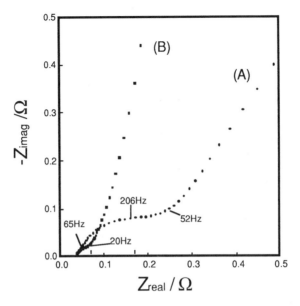

Fig.6   Impedance spectra for polyaniline film electrode before(A) and after(B) contact with DMcT in 0.1 M LiClO₄/AN.

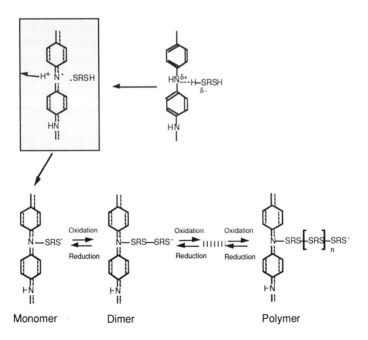

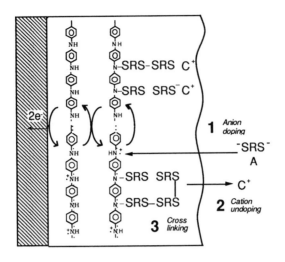

1: **Mass Increase**

2 &3: **Mass Decrease**

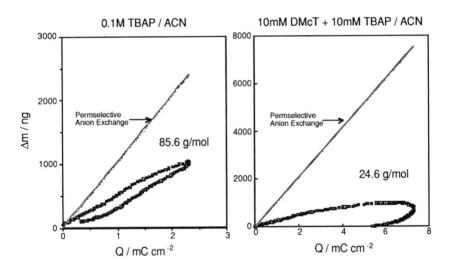

Fig.8　QCM responses(frequency vs. potential curves) for polyaniline electrode in AN solution containing (A) 10mM TBAP and (B) (10mM DMcT + 10mM TBAP).

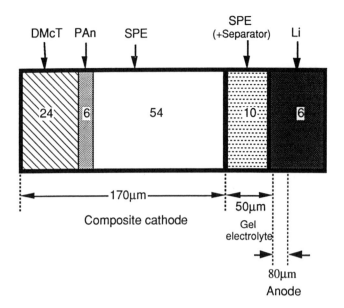

Fig.9   Cell configuration of Li/SPE/(PAn+DMcT) battery.

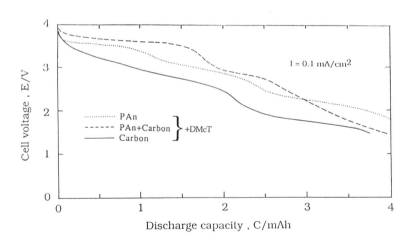

Fig.10   Typical discharging curves for the Li/SPE/(DMcT+PAn) cell described in Fig.9.

LITHIUM ROCKING CHAIR BATTERIES: AN OLD CONCEPT?

Bruno Scrosati
Dipartimento di Chimica, Università di Roma "La Sapienza"
Rome, Italy

ABSTRACT

In recent years consistent attention has been devoted to novel types of lithium rechargeable batteries in which the metal anode is replaced by a lithium-source anode. The general interest in these batteries, often called *rocking-chair batteries*, has increased consistently; however, the idea of exploiting the rocking lithium systems for achieving improvements in safety and cycle life is not new, dating back to the beginning of the eighties. In this paper we critically review the progress in the area and discuss the effective impact that rocking chair systems may have on the future of lithium battery technology.

THE ROCKING CHAIR CONCEPT

Primary lithium batteries now occupy an established role in the market and various types, having different sizes and capacities, are currently under commercial production in Japan, the United States and Europe. On the contrary, not even one example of a rechargeable lithium battery is today commercially available, despite the facts that: i) many electrodic systems have been proved to have a basic reversible behaviour and ii) a rechargeable lithium battery would be highly welcome as an alternative power source to conventional systems (e.g. nickel-cadmium), both in terms of energy density and of environmental control.

The main reason for the commercial failure of rechargeable lithium batteries lies in the reactivity of the alkali metal and in the related passivation and instability phenomena at its interface. Passivation limits the cyclability of the entire battery while instability, especially when caused by abuse or malfunctioning operations, may lead to a serious safety hazard. Indeed, a few incidents have occured, with occasional fires in equipment powered by lithium batteries, even for prototypes assembled in industrial laboratories having recognized experience in the battery technology.

One way of overcoming the problem could be the replacement of lithium by a non-metal compound, say Li_yMnYm, capable of storing and exchanging large quantity of lithium ions. In this way, rather than lithium plating and stripping as in the conventional systems, the electrochemical process at the negative side would be the uptake of lithium ions during charge and their release during discharge. If another non-metal Li-accepting compound, say $AzBw$, is used at the positive side, the entire electrochemical process would then involve the cyclic transfer of x equivalents of lithium ions between the two electrodes:

$$Li_yMnYm + AzBw \underset{\text{charge}}{\overset{\text{discharge}}{\rightleftharpoons}} Li_{(y-x)}MnYm + Li_xAzBw \qquad [1]$$

Therefore, these electrochemical systems may be described as concentration cells where the lithium ions 'rock' from one side to the other (Figure 1). Accordingly, these cells have been termed "rocking chair batteries".

Considering the nature of the electrochemical driving process, a successful operation for a rocking chair battery and its effective competion with pure lithium system would require some crucial conditions, namely:

i)   The lithium activity in the negative electrode LiyMnYm must be close to 1, in order to assure values of open circuit voltages approaching those obtainable with the pure lithium metal;

ii)   the equivalent weight of both electrodes must be low in order to assure specific capacity values of practical interest;

iii)   the diffusion coefficient of Li^+ ions in the ion-source Li_yMnYm negative electrode must be high in order to assure fast charge and discharge rates;

iv)   the changes in lithium ion chemical potential in both electrodes must be small in order to limit voltage variations upon Li^+ uptake and removal, and thus finally assure good voltage regulation upon charge and discharge;

v)   the ion-source compound must be easy to fabricate and based on non-toxic compounds, in order to assure low cost and enviromental control.

If all these conditions are satisfactorily met, the rocking chair concept may become of definite practical interest since it does offer concrete potential for extended cycling and safety. In fact, by in principle avoiding any metal plating process and thus eliminating the conditions for the growth of irregular or dendritic lithium, the chances of shorting and of overheating the battery system may indeed be considerably reduced.

HISTORY OF ROCKING CHAIR BATTERIES

The term "rocking chair battery" for lithium concentration cells of the type described above was, to our knowledge, first introduced by M. Armand in 1980 (1). The feasibility of the concept for the realization of long-cyclable lithium batteries was demonstrated by us in the early eighties (2-5) by assembling and testing small laboratory cell prototypes. In these cells, the lithium source negative electrode (i.e. the Li_yMnYm compound of equation [1]) was $LiWO_2$ or $Li_6Fe_2O_3$, respectively, and the non-intercalated, pristine, positive electrode (i.e. the A_ZB_W compounds of equation [1]) was TiS_2, WO_3, NbS_2 or V_2O_5, respectively. The electrolyte was a typical organic electrolyte, e.g. a solution of lithium perchlorate ($LiClO_4$) in propylene carbonate (PC).

These cells, which were based on electrochemical reactions of type [1], namely the reversible transport of lithium ions from one electrode to the other, had a reasonably high open circuit voltage and good charge-discharge efficiency upon extended cycling. Although not competitive with conventional lithium systems, the performance of these batteries clearly demonstrated that the rocking chair concept could indeed be exploited in practical devices and this attracted the interest of different authors from various laboratories. In 1987, Semkov and Sammels (6) reported on a rocking chair cell based on similar electrodic couples, but differing in that the $LiClO_4$-PC liquid electrolyte was replaced by a polyphosphazene-$LiCF_3SO_3$ solid polymer electrolyte (SPE). The cell had an open circuit voltage of 2.1-2.2 V and, due to the well known temperature limits of the conductivity of the polymer electrolyte(7), it could operate only at 90 C and at very low rates.

71

In 1987, Auburn and Barberio (8) investigated MoO_2 /$LiPF_6$-PC/ $LiCoO_2$ lithium concentration cells where the electrolyte was a solution of lithium hexafluorophosphate ($LiPF_6$) in propylene carbonate (PC). In these rocking chair systems the excess of cyclable lithium was directly provided by the native, lithium-rich compounds at the cathodic side.

The interest in rocking chair batteries remained moderate, with occasional reports and publications(9-12), up to most recent times when the concept gained substantial renewed attention following the announcement that a major Japanese Battery Company had reached the decision to exploit the idea for the large scale production of 'new design' rechargeable lithium batteries (13). Actually, this design is again based on Auborn and Barberio's idea of using a lithium-source cathode and a lithium-accepting anode combined in a rocking-chair system. The novelty lies in the choice of the anode material , in this case petroleum coke, Li_xC_6. The basic cell structure is then of the type:

$$Li_xC_6 \text{ / LiX in PC-EC / } Li_{(1-x)}CoO_2 \text{ (or } Li_{(1-x)}YO_2) \qquad [2]$$

where LiX is a lithium salt (e.g. $LiClO_4$), PC-EC is a mixed propylene carbonate (PC) - ethylene carbonate (EC) solvent and Y is a transition metal (e.g. Ni or Mn).

As expected, the electrochemical process is the transfer of lithium ions across the cell:

$$Li_yC + Li_{(1-y)}CoO_2 \underset{\text{discharge}}{\overset{\text{charge}}{\rightleftharpoons}} Li_{(y+x)}C + Li_{\{1-(y+x)\}} CoO_2 \qquad [3]$$

where y is about 0.2 Faraday per mole and the cyclable charge x is around 0.5 Faraday per mole.

The Japanese authors (13) named this cell 'lithium ion rechargeable battery', probably with the aim of also gaining some originality in terminology. However, as already pointed out, the system typically exploits the rocking chair approach and thus can be appropriately classified as an another member the rocking chair battery family. Undoubtedly, credit should be given to the Japanese Laboratory for the introduction of carbon as a novel negative electrode, an innovation which triggered substantial excitement in the field, since it gives expectations of improvements in cyclability, energy density and cost. In fact, according to the authors (13), the use of this anode material may lead to the development of D-size rocking chair batteries featuring a cycle life of 1,200 deep cycles and an energy density over three times that of nickel cadmium batteries.

Although not all the properties of the carbon anode are outstanding (see following discussion), it is not surprising that the presentation of the Japanese laboratory has stimulated the interest of the lithium battery community and that many research groups find the rocking chair idea appealing with a consequent new waive of publications in the field.

In 1991 Dahn et. al.(14) described the characteristics of a coin-size cell having the basic structure:

$$Li_xC_6 \text{ / } LiN(CF_3SO_2)_2 \text{ EC-DME / } Li_{(1-x)}NiO_2 \qquad [4]$$

which is essentially similar to that of the above described Japanese battery. In the Dahn et. al. work, the electrolyte is a 1M solution of $LiN(CF_3SO_2)_2$ in a 50/50 by volume blend of ethylene carbonate (EC) - dimetoxyethane (DME) and the cathode is

LiNiO$_2$, preferred to LiCoO$_2$ because of its lower voltage versus lithium and thus for its lower tendency to oxidize the organic electrolyte (see following sections).

Also in 1991, Tarascon and Guyomard(15) considered carbon-based rocking chair configurations using Li$_{(1+x)}$Mn$_2$O$_4$ as the lithium-source cathode material. The advantage is that the selected spinel compound is capable of providing an excess of lithium over that required for the overall electrochemical process, this accounting for the loss in capacity unevitably experienced by the carbon electrode during the initial irreversible uptake of lithium (16). The rocking chair cell studied by Tarascon and Guyomard(15):

$$Li_xC_6 / 1MLiClO_4 \text{ in } 50/50 \text{ EC-DME} / Li_{(1+x)}Mn_2O_4 \qquad [5]$$

presented a promising cycling behavior, in part, however, contrasted by poor voltage regulation and by a pronounced voltage drop at mid-composition.

In a 1991 patent (17), Slane and Plichta described a rechargeable lithium rocking chair battery of the type:

$$TiS_2 / LiAsF_6 \text{ in metyl acetate} / LiCoO_2 \qquad [6]$$

where the interesting aspect is that TiS$_2$, which is one the commonest *cathodes* in conventional lithium batteries, is here used as *anode* to accept Li$^+$ ions from LiCoO$_2$ during the charge process:

$$TiS_2 + LiCoO_2 \underset{\text{discharge}}{\overset{\text{charge}}{\rightleftharpoons}} Li_xTiS_2 + Li_{(1-x)}CoO_2 \qquad [7]$$

Finally, the current interest in the lithium rocking chair systems is demonstrated by the large numer of papers in the area presented at the 1991 Fall Meeting of the Electrochemical Society(18).

EXPECTATIONS AND LIMITS OF ROCKING CHAIR BATTERIES.

As already pointed out, the present enthusiasm for rocking chair batteries is based on the important expectation that the use of the Li$^+$ ion transfer electrodes may lead to consistent improvements in the safety and in the cycle life of lithium rechargeable systems. Indeed, should these expectations be fulfilled, rocking chair lithium batteries could shortly become viable power sources for a series of widespread and popular devices in the field of consumer electronics and appliances. However, with the possible exception of the Japanese case, lithium rocking chair batteries have not yet been proposed for commercial scale development. The explanation may lie in the fact that the penalities to be paid in driving the present design rocking chair systems, may still be too high to assure an effective competition with the conventional lithium systems or even more with the commercially established nickel- cadmium systems. Furthermore, it has to be pointed out here that the safety of operation, which has been always claimed as the most important inherent advantage of the rocking chair batteries, it is an assumption which still remain to be demonstrated in practice.

Therefore, the key question is to what extent the rocking chair innovation would be capable of affecting the progress of the lithium battery technology. In an attempt to provide some evaluation criteria, we discuss and analyze here some aspects of the basic characteristics and behavior of the rocking chair electrochemical systems.

Open Circuit voltage and voltage regulation.

One of the main requirements for selecting the proper Li-insertion compound to replace Li metal as the negative electrode in the rocking chair battery configuration, is that the Li^+ activity in the insertion compound should be as close to 1 as possible. In fact, only under this condition, can the selected compound exhibit a potential approaching that of the metal and thus, ultimately, give rocking chair batteries having open circuit voltages approaching those of the conventional lithium batteries. In the past the choices have been lithium-rich iron (5,9) or tungsten (8,11) oxides and, presently, petroleum coke (13-15).

Figure 2 compares the potential ranges of various Li-insertion compounds. The potentials are relative to the Li metal and reported for the exchange of x numbers of Li equivalents per mole. The figure indicates that the choice of Li_xC_6 appears appropriate from the point of view of the potential value, which ideed approaches that of Li metal in the fully-intercalated state (x=0.5).

However, the value of the potential is not the sole criterion of importance in the selection of the Li-insertion anode; the variation of potential upon Li^+ exchange is also a crucial factor. Indeed, it desirable to have a rocking chair battery which is not only capable of exhibiting a high open circuit voltage but also a good voltage regulation upon charge and discharge. In this respect the choice of petroleum coke, Li_xC_6, anode remains questionable, since its potential varies as much as 1.5 V upon the exchange of the total removable lithium (0<x<0.5) (14).

Therefore, the poor voltage regulation drawback outlined at the initial stage of the rocking chair development (3), still remains a problem even in the case of the recent generation systems. Indeed, this can be a serious drawback, especially in view of applications in the electronics field where voltage stability is often a priority requirement. Future attention to the rocking chair battery development should then be devoted to the characterization of anode materials having not only a lithium activity approching one in the lithium-rich state, but also a limited change in Li^+ chemical potential upon lithium exchange. In this respect the choice should be oriented towards compounds having open structures with loose interactions with the hosted lithium ions.

Specific capacity and energy density.

As is well known, lithium metal is almost an ideal electrode material due to its high specific capacity which amounts to 3.86 Ah/g. The substitution of Li by any Li-inserted compound does necessarily lead to severe sacrificies in capacity content. For instance, in the case of the Li_xC_6 (x=0.5) choice, the specific capacity drops of more than one order of magnitude (e.g. 0.18 Ah/g). Even more dramatic is the loss associated with other proposed anode compounds, such as $Li_{(6-x)}Fe_2O_3$ or Li_xWO_2.

These losses in capacity are directly reflected in losses in energy density. Figure 3 shows values of theoretical energy density (in Wh/kg) related to combinations between a Li_xC_6 (x=0.5) anode and various cathode materials. In the same figure the theoretical energy density of a typical combination in 'conventional' lithium batteries, i.e. that of the Li/V_6O_{13} couple, is also shown for comparison purposes.

It may be clearly seen that even under the best conditions (i.e. using the Li_xC_6 anode in combination with the $LiCoO_2$ cathode) the maximum achievable energy density is at least 50% lower than that of comparable systems based on Li metal anodes. Unfortunately, there are not practical ways to overcome this drawback since it is inherent in the choice of anode materials alternative to lithium metal. Therefore, if priority is given to rocking chair batteries, one unevitably has to deal with comparatively low energy density systems and cope with the related penalities by enhancing other counterbalancing advantages specific to the rocking concept.

Rate capability.

The electrochemical process in rocking chair batteries is the exchange of Li^+ ions between the two electrodes. Accordingly, the diffusion kinetics of the Li^+ insertion-deinsertion process directly control the rate of charge and discharge of these batteries. Since it is reasonable to assume that the diffusion of Li^+ ions throughout the structure of most of the solid host compounds which are proposed as electrodes in rocking chair batteries is slow, rate limitation may indeed be matter of concern in the utilization of these batteries. Fortunately, this is not a thermodynamic but, rather, a kinetic limitation and thus the problem can be somewhat solved, or at least alleviated, by selecting proper electrode compositions and morphologies, such as those involving thin-film, large surface area, configurations.

Reversibility and cyclability.

Extended cyclability of the rocking chair systems requires high reversibility of the electrode processes. Since one side electrode provides the lithium for the other side electrode, the amount of cycled charge is directly correlated. Consequently, even if one electrode has an efficiency as high as 99% (which is itself a difficult target to achieve in practice), the life of the cell is still limited to 100 cycles.
Furthermore, it has been ascertained that in systems using the Li_xC_6 anodes, 20% of the total capacity is unevitably lost due to the partial irreversibility of the initial intercalation process of Li^+ ions into the pristine carbon structure (15,16).
These combined facts readily indicate that a successful operation of Li_xC_6 anode rocking chair batteries requires the use of electrode materials cabable of providing a large excess of lithium with respect to the faradaic needs. In this connection, Li-insertion compounds characterized by reversible processes involving more than 1 Li equivalent per mole, should be preferred.

Safety.

The major issue which has backed the renewed interest in rocking chair batteries is certainly the expectation of a safe operation . Indeed, this is a very valid issue since it is reasonable to expect that the replacement of the lithium metal with a lithium-intercalated carbon electrode would considerably reduce the the chances of ignition of those uncontrolled side reactions and processes which are responsible for the failure and the incidents experienced in lithium batteries.
However, one must be aware of the fact that also in the case of rocking chair configurations deviations from the expected electrochemical process cannot be totally excluded. Indeed, depending on the type of negative electrode and on the cycling regime, conditions for promoting lithium plating rather than lithium insertion may unintentionally be created. For instance, in the usual example of systems based on the Li_xC_6 anode, one can reasonably consider that uncontrolled operation conditions

may drive the polarization of the carbon electrode to values at which lithium plating may become the predominant process. In fact, these conditions may very likely occur in the case where the cell is anode limited: attempts to pull out the residual lithium stored at the cathode side will unevitably drive the potential of the carbon electrode to regions favourable for Li plating, with the risk of building up regions of highly reactive metallic lithium deposited on a finely subdivided carbon surface.

Another point of concern in establishing the safety of the operation of rocking chair batteries is related to the decomposition of the electrolytes, which in the common case of the liquid organic solutions averages around 4 V vs. Li (19). Electrolyte media having higher electrochemical stability than the liquid organic solutions, should be chosen to assure the safe operation of Li_xC_6 / $Li_xM_yO_z$ rocking chair batteries. In this respect, a possible approach is the replacement of the liquid electrolyte with a solid electrolyte. Promising candidates are the Li^+ conducting polymer electrolytes (7) which are on average characterized by electrochemical windows exceeding 4 V vs. Li (7) and, in certain cases, even wider than 5 V vs. Li (20).

The future.

As concluded in the previous sections, the development of rocking chair batteries capable of effectively replacing the more conventional lithium systems and the commercially established nickel cadmium systems, is still a goal to be achieved. The major drawbacks appear at the moment to be poor voltage regulation and low power outputs.

However, in the cases where the main requirement is a low-rate, reproducible and safe operation, the rocking-chair system may indeed be a winning concept even in the present design. An illustrative example is given by the electrochromic 'smart' windows which are essentially batteries, often lithium-based, where the energy output is revealed by color changes (21). All the most successful prototypes of these optical devices, some of them presently under industrial production as innovative car rear-view mirrors , use the rocking-chair concept. In fact, they are formed by a main Li^+-ion insertion, an electrochromic electrode (the positive electrode in the battery analogue) and a Li^+ ion-rich counter electrode (the negative electrode in the battery analogue) and the electrochemical process of these optical electrochemical devices is again the swing transfer of lithium from one side to the other of the cell.

In the electrochromic, rocking-chair systems the value of open circuit voltage and that of the energy density are scarcely of any importance with respect to long cyclability and to sharp optical response. Some concern could still be addressed to the charge-discharge rates since they directly influence the device's response time. However, for most of the applications, response times of the order of one or few seconds are appropriate and thus even materials with relatively slow lithium diffusion rates are still acceptable, providing they offer the required optical properties.

Therefore, the first generation rocking chair batteries, proposed in the early eighties as novel lithium, low-rate rechargeable systems capable of offering long cycle life (3), have indeed found a valid response in practice. So, the old concept does hold. The question is whether the new generation, new concept rocking chair batteries, proposed in alternative to high-energy, high-power lithium and nickel-cadmium batteries (13,14), will eventually find their way in the competitive portable power source market.

76

ACKNOWLEDGEMENT.
    The author wishes to thank Dr. B.B. Owens of the University of Minnesota for
providing data for calculations and for the helpful discussion.

REFERENCES

1.   M. Armand, in 'Materials for Advanced Batteries', D.W. Murphy,
     J. Broodhead and B.C.H. Steele Eds,Plenum Press, New York 1980, pag.145.
2.   F. Bonino, M. Lazzari, L. Peraldo Bicelli, B. Di Pietro and B.   Scrosati, in
     "Proc. Symp. on Lithium Batteries", H.V. Venkatasetty ed. The Electrochem.
     Soc. Pennington, N. Jersey 1981 pag. 255
3.   M. Lazzari and B. Scrosati,*J.Electrochem.Soc.* **127,**773 (1980).
4.   B. Di Pietro, M. Patriarca and B. Scrosati, *J. Power Sources*, **8,** 289 (1982).
5.   M. Lazzari and B. Scrosati, U.S. Patent 4,464,447 Aug. 7, 1984.
6.   K.W. Semkov and A.F. Sammels, *J.Electrochem.Soc.,* **134,**766 (1987).
7.   "Polymer Electrolyte Review I", J.MacCallum & C.A.Vincent Eds,
     Elsevier Applied Science Publishers, London, 1987.
8.   J.J. Auborn and Y.L. Barberio, I*J.Electrochem.Soc.,* **34,** 638   (1987)
9.   K.M. Abraham, D.M.Pasquariello, E.B.Willstaed and G.F. McAndrews,
     Proceedings Symp. "Primary and Secondary Ambient Temperature Lithium
     Batteries", J.P.Gabano,   Z.Takehara and P.Bro, Eds, Vol. 88-6, The
     Electrochem. Soc., Pennington, N.J., pag.668, 1988.
10. C.D. Dessjardine and G.C. MacLean, Fall Meeting Electrochem. Soc.,
     Hollywood, Fla, Oct. 15-20, 1989 , abstr. N. 52.
11. F.A.Uribe and A.F.Sammels, Proceedings Symp. "Materials and
     Processes for Lithium Batteries", K.M.Abraham &B.B.Owens, Eds,
     Vol. 89-4, The Electrochem. Soc.,Pennington, N.J., pag. 201, 1989.
12. R.S.MacMillan, Fall Meeting Electrochem.Soc., Seattle, Wa., Oct. 14-19,
     1990, abstr. N.43.
13. T. Nagaura and K. Tazawa, *Prog. Batteries Sol. Cells*, **9,** 20(1990).
14. J.R.Dahn, U.vonSacken, M.R.Jukow and H.Al-Janaby, *J.Electrochem.Soc.,*
     137 2207 (1991).
15. J.M.Tarascon and D.Guyomard, *J.Electrochem.Soc.,* **138,** 2864(1991).
16. R.Fong, U von Sacken and J.R. Dahn, *J.Electrochem.Soc.,***137,** 2009 (1990).
17. S.M.Slane and E.J.Plichita, US Patent 4,983,476, Jan. 8, 1991.
18. B. Scrosati, Fall Meeting Electrochem.Soc., Phoenix, Arizona   Oct.13-17,
     1991, abstr. N. 31; N.,Imanishi, S.Ohashi, Y.Takeda, O.Yamamoto and M.
     Inakaki, ibid, abstr. N.32; M.E.Bolster, ibid, abstr. N. 33; Y.Matsuda,
     M.Morita, T.Hanada and M. Kawaguchi,   ibid, abstr.N. 35; D.Guyomard
     and J.M,Tarascon, ibid, abstr.36.
19. D.Aubarch, M.Daroux, P.Faguy and E.Yeager, *J. Electroanal, Chem,*   **297,**
     225 (1991.
20. K.M. Abraham, M.Alamgir, G.S.Jones and L.L.Wu, Extended   Abracts,
     Electrochemical Society , Vol. 91-2, Fall Meeting, Phoenix, Arizona, Oct.
     13-17, abstr. N.691, pag. 1029 (1991).
21. "Materials Science for Solar Energy Conversion Systems", C.G.Granqvist
     Ed., Pergamon Press,Oxford, 1991.

ROCKING CHAIR BATTERY

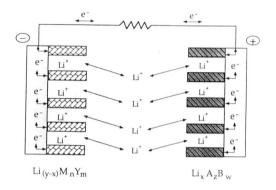

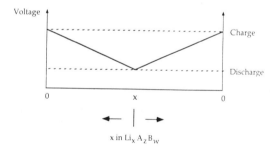

Figure 1-Model of a rocking chair battery and of the related charge- discharge voltage profile.

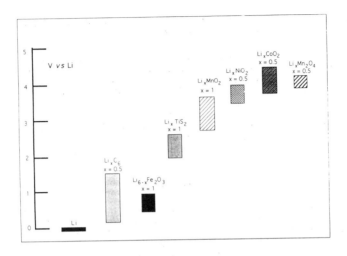

Figure 2-Potential ranges (versus Li) of various Li-insertion compounds, calculated for the exchange of x Li equivalents per mole.

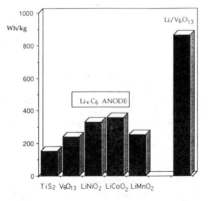

Figure 3-Theoretical energy density for electrodic systems based on the combination of Li_xC_6 (x=0.5) with various cathode materials. The theoretical energy density of the Li/V_6O_{13} couple is also shown for comparison purposes.

THE STRUCTURE AND CHARGE-DISCHARGE CHARACTERISTICS OF MESOPHASE-PITCH BASED CARBONS

N.Imanishi, H.Kashiwagi, Y.Takeda and O.Yamamoto
Department of Chemistry for Materials, Faculty of
Engineering, Mie University
1515 Kamihama-cho, Tsu, Mie 514, Japan

M.Inagaki
Department of Applied Chemistry, Faculty of
Engineering, Hokkaido University
North 13, West 8, Sapporo 060, Japan

ABSTRACT

4 types of highly oriented Mesophase-pitch
based carbons were investigated with respect to
electrochemical characteristics and structural
change. Differences in orientation effected on
the charge-discharge characteristics. The radial
oriented carbon fiber showed low polarization
and high capacity of lithium intercalation. But,
side-reaction also occurred much at the same
time. Onion-like orientation showed low capacity
of intercalation and side-reaction also hardly
occurred on surface. Low crystalline Mesophase-
pitch based carbons were also investigated but
did not show lithium intercalation and reversible
reaction.

INTRODUCTION

The study on the lithium anode is important for
development of practical lithium secondary battery.
Lithium metal does not work enough as reversible electrode,
because lithium reacts with electrolytes(1). Much efforts
were made to improve the characteristics of lithium anode
such as searching suitable electrolyte and making alloy
with other metals(2,3). Another possible way to prevent
this reaction is using other materials as a reversible
electrode. One of most widely studied material is the
carbon(4,5). Many researchers reported lithium can
intercalate and deintercalate from carbon reversibly.
However, carbon have many variety of structures. For
example, there are d-values, crystallite size, orientation,
surface area and so on. They are different in each carbons
which were prepared by different method. These parameters
have to be controlled in order to discuss on
the electrochemical character of the carbon. In this

study, effects of orientation was focused, and the charge-discharge behaviors were investigated.

EXPERIMENTALS

Mesophase-pitch based carbons annealed at high temperatures(2600 and 2800 °C) and low temperature(1200 °C) were used. They were supplied from prof. Inagaki of Hokkaido University. There are four different orientations in each group. For high temperature carbons, their orientations are illustrated as radial with wedge, radial, random and onion-like structure. The cross-sectional view was shown in Fig. 1. The first one was very easy to cleave along the radial plane. This is resulting from highly oriented structure. They were denoted by simply I-1, I-2, I-3 and I-4. We can expect I-1 and I-2 are suitable orientations for lithium intercalation and I-4 works as blocking electrode. Diameter of the fiber was about 10 μm. The same four types of carbons annealed at 1200 °C were also used. They are amorphous under X-ray diffraction analysis. Differences in orientations can not be observed clearly as high temperature carbons.

Figure 2 shows two kinds of experimental cells. The three electrode cell was used to measure exactly charge-discharge behavior. Reference and counter electrodes were lithium sheets. Carbon fibers were cut into 5 mm length and the weight of them were measured carefully by microbalance. The fiber was held between nickel plates and soaked into the electrolyte. The electrolytes was 1 M LiClO$_4$ dissolved into PC-THF mixed solvent. Excess amount of solvent was always used to get rid of the effects of composition change of electrolytes. The another one is coin type cell for X-ray analysis. A large amount of sample were needed for X-ray measurements. The electrolyte was 1 M LiClO$_4$ in PC-THF mixed solvent. The X-ray measurement were performed under Ar atmosphere.

RESULTS AND DISCUSSION

At first, the electrochemical characteristics for each carbon were surveyed by using cyclic voltammogram. Figure 3 shows results of high temperature carbons. 1 and 2 show broad cathode and anode peak around 0.5 V. These peaks are seemed to be lithium intercalation and deintercalation. 3 and 4 show only small peaks around 0 V, which are probably corresponding to lithium dissolution and deposition. It is concluded that these differences are coming from their difference in crystal orientation. Figure 4 shows results of low temperature carbons. They show very similar behaviors each other. The peaks around 0 V are corresponding to lithium dissolution and deposition. The mesophase-pitch based carbons with low crystallinity are not so active electrochemically as high temperature ones.

Next we examine the effects of crystal orientation by using three electrode cell as shown before. Figure 5(a) shows the charge-discharge curves of I-1 of high temperature carbons. The current density was 30 μAmg^{-1}. The cut off voltages are 0 V and 3.5 V. This carbon shows long plateau at around 0.8 V. And the intercalation seems not to occur because charge capacity is almost zero. This is very contradictory fact because radial structure have to be suitable for lithium intercalation. However, this results shows most of the capacity is irreversible reaction such as electrolyte decomposition or destruction of the structure. We observed the discharged carbon fiber by Scanning electron micrographs. Figure 6(a) shows the situation of the surface before and after discharge. Apparently the fiber no longer keep the basic structure, because this type of carbon is very easy to cleave as mentioned before. And some unknown products covered the surface of the carbon. It is concluded that these products may be formed at around 0.8 V and cause the cleavage of fiber. The reversibility was lost in this discharge plateau.

Figure 5(b) shows the results of I-2 carbon. I-2 also has radial orientation structure, but does not cleave easily like I-1. The discharge profile shows normal lithium intercalation which was indicated by potential plateau at ca. 100 mV. The charge capacity was about 300 $mAhg^{-1}$ which was close to the theoretical capacity. These behavior is resulting from radial and more stable structure than I-1. SEM observations for this products were shown in Fig 6(b). There are no large differences between before and after discharge. The fiber did not cleave like I-1 carbon even after discharge. However, some products covered on the surface of the carbon as same as I-1. This is probably produced during early period of discharge, because the 0.8 V plateau of I-1 carbon indicates the formation of by-products.

Figure 5(c) shows the results of carbon I-4 with onion like orientation. This carbon shows most simple charge-discharge curves. The voltage dropped much more sharply just after starting discharge than former two carbons. I-1 and I-2 show the occurrence of some products on the surface during these early period. And this behavior predicts side-reaction hardly occur through the discharge of I-4. The charge capacity was almost as same as discharge capacity. This means good reversibility, but the problem is low capacity itself. However, other point is ideal as anode of lithium battery. SEM observations were also performed for discharge products as shown in Fig. 6(c). In this case, the destruction of structure did not be observed. And any products could not be observed on the surface of carbon fiber. We can concluded that I-4 has small capacity of lithium intercalation, but side-reaction also hardly occur for the I-4. This is suggesting the reaction point which concerns with the by-products.

We also examined low temperature carbons as shown in Fig. 7. The solid line shows 1-2 and doted line shows I-4 at 1200 °C. The discharge curves are different from high temperature carbons. The potential decreased linearly to 0 V and did not show the plateau at around 100 mV which corresponds lithium intercalation. This means that the normal lithium intercalation no longer occurred in these carbons. Comparatively higher voltage part in early period was considered to be side-reaction such as decomposition of electrolyte. Cyclability was lower than higher carbon. It is concluded that these carbons might not have the sites which can accommodate lithium ion and can not be used as anode.

We examined the structural change during the discharge process. The samples were discharged potentiostaticaly and the circuit was kept opened for one day before the X-ray analysis. The potentiostatic discharge were proceeded until the current through the cell decreased below $5\mu A$. Figure 8 shows results of I-2 radial carbon. The (002) peak corresponding to the basal plane moved to the lower angles through discharge. This means lithium intercalation occurred below ca. 0.5 V. Figure 9 shows results of I-4 at high temperature. This carbon was recognized as blocking electrode for lithium intercalation. However, it also shows the expansion of the structure along c-axis. This means the lithium intercalation practically occurred in this carbon. We can compare these two results at 0.5 V, which shows that apparently movement of (002) peak was faster in I-2 than I-4. This means lithium intercalation in I-2 carbon occur much easier kinetically. The effect of structure appeared in the polarization behavior which concern with the discharge capacity of each carbon.

The relation between discharge behavior and structural change can be described from these results. From 3.0 to 0.5 V, lattice expansion cannot be observed. In this region, the other reaction may occur as stated before. The intercalation started when potential reached below 0.5 V and the lattice began to expand along the c-axis. The expansion stopped when d-value increased about 7.1 A. This value was attained when the capacity was almost corresponding to C_6Li.

CONCLUSION

All the results can be summarized as follows. Carbon with low crystallinity did not show the capacity of lithium intercalation. Highly oriented I-1 at 2800 °C shows cleavage of the structure through the discharge. And other reaction except for intercalation occurred at ca. 0.8 V. I-2 at 2800 °C showed good characteristics. The charge-discharge capacity was about 300 mAhg^{-1} and it showed good reversibility. Side-reaction hardly occur on the surface of I-4. However, the capacity itself was lower. From the results of high temperature three carbons, we can conclude that not only lithium intercalation, but the side-reaction can be influenced by the orientation of carbon crystallite.

REFERENCES

1. E.Peled, in "Lithium Batteries," J.P.Gabano, Editor, 43, Academic Press, London(1983).

2. C.D.Desjardins, G.K.Maclean, and H.Sharifian, J.Electrochem.Soc., 136, 345(1989).

3. D.Aurbach, Y.Gofer, and J.Langzam, ibid., 136, 3198 (1989).

4. R.Fong, U.v.Sacken, and J.R.Dahn, ibid., 137, 2009(1990).

5. J.R.Dahn, U.v.Sacken, M.W.Juzkow, and H.Al-Janaby, ibid., 138, 2207(1991).

I-1   I-2

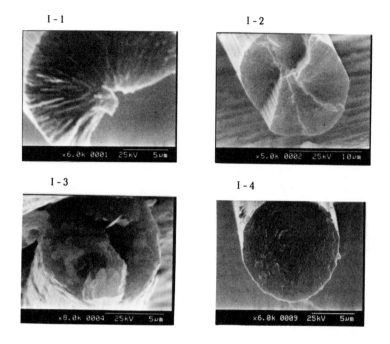

I-3   I-4

Fig.1 The cross-sectional view of Mesophase-pitch based
      carbons.

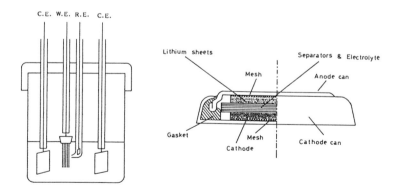

Fig.2 The construction of experimental cells.

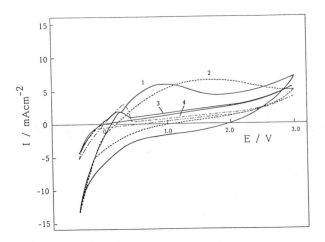

Fig.3 The cyclic voltammograms of carbon fiber(2800°C).

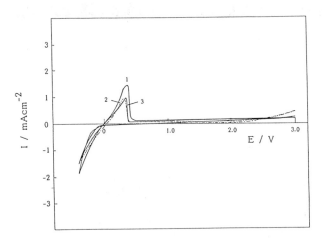

Fig.4 The cyclic voltammograms of carbon fiber(1200°C).

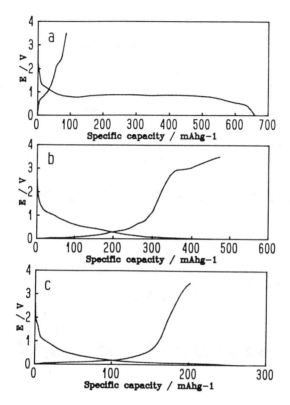

Fig.5 The charge-discharge curves in PC/THF electrolyte.
a) I-1 b) I-2 c) I-4

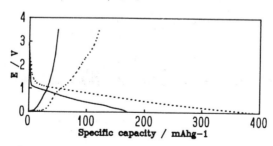

Fig.7 The charge-discharge curves of carbon fiber(1200°C).
solid line:I-2   doted line:I-4

BEFORE   AFTER

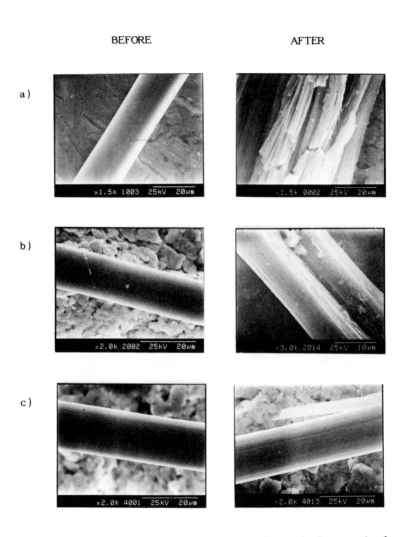

Fig.6 The SEM observations of carbon fiber before and after
    discharge. a) I-1 b) I-2 c) I-4

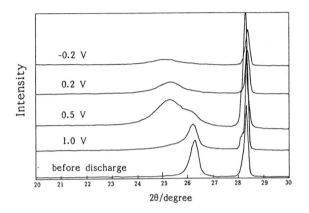

Fig.8 The X-ray diffraction patterns of I-2(2800°C)
at various discharge level.

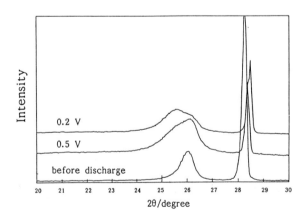

Fig.9 The X-ray diffraction patterns of I-4(2600°C)
at various discharge level.

89

A STUDY OF CARBONS AND GRAPHITES AS ANODES FOR
LITHIUM RECHARGEABLE CELLS

Mary Elizabeth Bolster

SAFT AMERICA INC
Research and Development Center
109 Beaver Court
Cockeysville, MD 21030

ABSTRACT

The carbon based, lithium ion "rocking chair"
anode has been suggested as a viable replacement
for elemental lithium in rechargeable cells. This
approach could solve safety problems related to
the reactivity of high surface area lithium and
dendrites formed during cycling. The lithium
cycling characteristics of carbon blacks,
graphite, coke, PAN foam and fibers, and mesophase
pitch fibers were studied to determine the type of
carbons that would be suitable for use. Coke,
PAN, and pitch materials demonstrated reversible
lithium ion intercalation. The PAN fibers were
found to possess the highest capacity and charge
retention. The carbon structure and its degree of
graphitization was indicated as the most important
factor that decides the nature and extent of the
lithium intercalation process.

INTRODUCTION

There is increasing interest in lithium ion anodes to
replace elemental lithium in high energy density
rechargeable systems. Lithium rechargeables presently have
limited commercial applicability due to the reactivity of
lithium. The carbon based "rocking chair" lithium
intercalation anode has received considerable attention in
the scientific community since the reported development of a
3.6 volt lithium ion rechargeable battery using a
Carbon/CoO_2 electrochemical couple (1,2). Superior cycle
life and charge retention was claimed compared to the

present 1.5 volt Ni-Cd system. It is known that graphites are able to intercalate lithium due to their layered lattice structure (3). These factors have stimulated the investigation of carbonaceous materials as lithium ion anodes (4, 5, 6).

A voltage level during discharge and charge of 0.5 volts or less is desirable to preserve the advantageous high potential of metal oxide cathodes such as MnO_2, V_2O_5, and CoO_2. The material must also possess a high theoretical capacity to minimize the weight of carbon that needs to be incorporated into the practical cell. Cell capacities equivalent to that offered by a lithium anode cell must be realized to be a practical alternative. Also, relatively rate independent voltage and capacity, as well as good cycle life and charge retention must be demonstrated.

In this work, the electrochemical behavior of carbon blacks, graphite, polyacrylonitrile (PAN) foam and fibers, and mesophase pitch fibers were studied by constant current cycling with lithium. A fixed current density of 1 mA/cm^2 was employed on flat plate electrodes with excess lithium capacity and in an electrolyte flooded condition. The voltages under discharge and charge were observed, and the mAh/g capacity for lithium was correlated for the carbon sample types.

EXPERIMENTAL

The carbon anodes were prepared by forming teflon bonded electrodes on nickel exmet current collectors. Carbon fibers were received prechopped, and the PAN foam sample was powderized. When the fiber's adhesion to the exmet was relatively poor, they were sandwiched within a folded section of exmet. The lithium electrode also utilized nickel exmet as the collector, and nickel leads were used for both electrodes. Each electrode was precut to dimensions of 2 x 2 cm^2. A simple glass cell was used for the electrochemical cycling tests. The cell was assembled with the lithium and carbon flag electrodes placed flat at the base of the cell. Celanese Celgard 2400 microporous polypropylene separator enclosed the lithium electrode. The cell was filled with a 10cm^3 excess of 1.0M $LiAsF_6$ in 1:1 PC/DME electrolyte. Weight was continuously applied to the cell during the tests to ensure good interfacial contact. All experiments were conducted in an argon filled drybox. Either an EG&G PAR 273A Potentiostat Galvanostat or a LAMDA power supply in conjunction with a Kikusui electronic load device were used to maintain constant current control. A current of 4 mA was applied to the $4cm^2$ area cell, corresponding to 1 mA/cm^2 on each single electrode face.

An open circuit of 0.5 hour minimum was allowed prior to cycling. Primary discharge versus lithium, or lithium intercalation, was continued until the cell reached zero volts. The charge, or deintercalation, was carried out for an equivalent time period corresponding to the discharge. A minimum rest time of 0.5 hour was also maintained between the forward and reverse steps. Usually one cycle is sufficient to determine if reversible lithium intercalation was present, however, several cycles were performed on reversible types to see if the behavior was repetitious. Each sample was tested in duplicate to ensure the validity of the results. The carbon weights reported are corrected for the weight of binder.

RESULTS AND DISCUSSION

The carbon blacks and graphite demonstrated irreversible behavior (Table 1). Typically, the discharge/charge profiles were as shown for Shawinigan acetylene black in Figure 1. A voltage plateau of approximately 1.0 volt was observed on primary discharge versus lithium, that would correspond to lithium ion incorporation. The majority of the discharge in carbon blacks and graphite occurred before 0.5 volt was reached. On the charge, or lithium deintercalation step, there was no evidence of reversible, solid state lithium ion intercalation by a voltage region corresponding to the 1.0 volt discharge. The cells immediately progressed to the solvent oxidation region at 4.5 volts. There was some correlation of capacity to zero volts with carbon surface area, although Cabot Black Pearls 2000 and the graphite fibrils were an exception. The BP-2000 was lower in capacity, and the graphite fibrils higher than their relative surface areas suggest. For instance, Shawinigan black has a surface area of only 67 m^2/g and gave 381 mAh/g to zero volts. Ketjenblack EC600 has a surface area of 1250 m^2/g and gave 933 mAh/g to zero volts. However, BP-2000, with 1475 m^2/g area, resulted in only 165 mAh/g, and graphite fibrils, with a lower surface area of 250 m^2/g, had a very high capacity to zero volts of 1783 mAh/g. The stoichiometry for lithium ion is given in Table 1, although the predominate process is probably the reduction of propylene carbonate on the carbon surface, to form insoluble lithium carbonate, or alkyl carbonates within the carbon matrix. The greater the available surface sites, the longer this electrochemical reduction can occur before complete saturation is reached. Carbon blacks have an amorphous, disorganized structure that would disfavor ordered solid state lithium ion intercalation. Graphite has well defined lattice spacing, but in this effort was found to demonstrate the same irreversible behavior. The spacing available is not appropriate to allow the solvated lithium ion to enter without loss of structural integrity.

Three types of petroleum needle coke were examined. One was a commercially available calcined needle coke, graphitized at 1500 $^{\circ}$C. This coke was also partially graphitized at 2250 $^{\circ}$C and fully graphitized at 3000 $^{\circ}$C (Table 2). The calcined coke first and second cycles are represented in Figures 2 and 3. Low voltage reversibility for lithium ion intercalation was demonstrated. The higher temperature graphitizations (Figures 4 and 5) resulted in irreversibility similar to the carbon blacks. The calcined PET coke showed a lower voltage on discharge, averaging 0.5 volt at 50% depth of discharge. On charge, deintercalation occurred from 0.5 to 1.5 volts. Subsequent to the removal of the lithium ion, the cell rose to the solvent oxidation plateau region. This particular flat plate cell at the 1 mA/cm^2 rate gave only 66 mAh/g to zero volts and 38 mAh/g reversible capacity. These cells contained a loading of 0.125g/cm^2 in a relatively thick plate of 0.76mm. Perhaps the performance could be improved for these rates by lamination of very thin electrodes. Petroleum coke has mesophase characteristics due to the alignment of the graphite planes as the material is coalesced from a melt, thus forming a needle like structure. These linear planes have minimal cross bonding allowing the apparent diffusion of a solvated Li+ species without structural disintegration and solvent entrapment.

Three sources of PAN material were selected. Two were fibers, obtained from AMOCO and Courtaulds Grafil. The foam was obtained from EG&G and was powderized to form a TFE bonded electrode. Also, two mesophase pitch fibers were obtained from AMOCO. The results are shown in Table 3. The AMOCO T-50 is a high modulus fiber and gave only 29 mAh/g discharge capacity and 25 mAh/g charge capacity, although the charge retention was higher than for the coke carbon. The two Grafil fibers resulted in the best performance. The XA-S fiber has a standard modulus and demonstrated 124 mAh/g on discharge and 76 mAh/g on charge at the 1 mA/cm^2 current density. The HM-S fiber has a high modulus and gave a higher charge capacity of 106 mAh/g. The characteristics of the first and third cycles are shown in Figures 6 and 7. The discharge and charge voltage of the PAN fibers are lower than for the coke carbon. Most of the discharge capacity is below 0.5 volt, and the lithium ion is removed at less than

1.0 volt. Again, the charge retention during OCV was higher than for the coke. PAN fibers possess concentric parallel graphite planes, formed during pyrolysis of the parent polymer. The ground PAN foam was essentially irreversible, giving the higher 1.0 volt discharge plateau and long discharge capacity typical of carbon blacks.

The P-55 mesophase high modulus pitch fiber gave capacities similar to coke (Figure 8). The voltage levels on discharge and charge were equivalent to the PAN fibers. The P-100 fiber essentially did not support either electrochemical discharge or charge. It is an ultra high modulus fiber with very high tensile strength and thermal conductivity characteristics. The mesophase pitch fibers are formed by extrusion of molten coal tar through an aperture, causing highly organized graphite planes that are radially aligned. The orientation of the alignment is a variable of spinning the fiber during the extrusion process.

CONCLUSIONS

Petroleum coke, PAN fibers, and certain types of mesophase pitch fibers were shown to have reversibility for lithium ion intercalation. The grafil PAN fibers demonstrated the lowest voltage, highest capacity per gram weight of carbon and excellent charge retention when cycled at a 1 mA/cm^2 current densityon flat plate electrodes. The degree of graphitization was demonstrated to play an important role in the performance of the coke carbon. A low temperature of about 1500 $^\circ$C is necessary to produce the desired reversible process.

REFERENCES

1.   SONY Enertec, February 14, 1990.
2.   R.T. Nagaura, J.E.C. Newsletter No. 2 (Mar-Apr), 17 (1991).
3   J.O. Besenhard and H.P. Fritz, Angew Chem Ed. Engl., 22, 950 (1983).
4.   R. Fong, U. von Sacken, and J.R. Dahn, J. Electrochem. Soc., 137, 2009 (1990).
5.   M. Mohri, N. Yanagisaua, Y. Tajimi, H. Tanaka, T. Mitate, S. Nakajimi, M. Yoshida, Y. Yoshimoto, T. Suzuki, and H. Wada, J. Power Sources, 26, 545 (1989).
6.   M. Arakawa and J.I. Yamaki, J. Electroanal Chem., 219, 273 (1987).

TABLE 1

Carbon Black Capacity at 1 mA/cm² in 1M LiAsF₆ in 50/50 Volume % PC/DME

Carbon Type	Weight (g)	Discharge Hrs to OV	Charge Hrs to 1.5V	Discharge Capacity (mAh/g)	Charge Capacity (mAh/g)	Discharge Formula	Charge Formula
Shawinigan	0.104	9.9	0.8	381	31	$Li_{1.02} C_6$	$Li_{0.08} C_6$
Ketjenblack EC600	0.060	13.0	0.0	933	0	$Li_{2.50} C_6$	$Li_0 C_6$
Black Pearls 2000	0.068	2.8	0.0	165	0	$Li_{0.40} C_6$	$Li_0 C_6$
Graphite Fibril	0.081	36.1	0.0	1783	0	$Li_{4.79} C_6$	$Li_0 C_6$
Lonza Graphite	0.264	1.8	0.0	27	0	$Li_{0.01} C_6$	$Li_0 C_6$

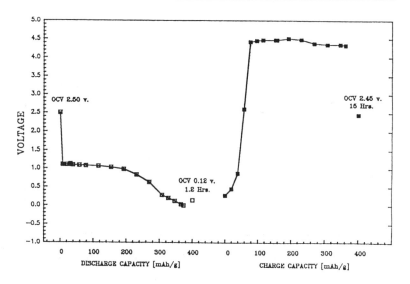

Figure 1: Shawinigan Acetylene Black
Voltage vs. Capacity (1st Cycle)

TABLE 2

PET Coke Capacity at 1 mA/cm² in 1M LiAsF₆ in 50/50 Volume % PC/DME

Carbon Type	Weight (g)	Discharge Hrs to OV	Charge Hrs to 1.5V	Discharge Capacity (mAh/g)	Charge Capacity (mAh/g)	Discharge Formula	Charge Formula
Calcined PET Coke	0.502	8.3	4.8	66	38.0	$Li_{0.18}C_6$	$Li_{0.10}C_6$
Partially Graph Coke	0.493	9.0	0.1	73	0.8	$Li_{0.20}C_6$	$Li_{0.02}C_6$
Fully Graph Coke	0.536	20.7	1.2	155	8.2	$Li_{0.42}C_6$	$Li_{0.02}C_6$

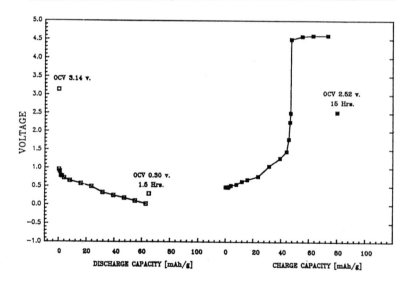

Figure 2: PET Coke - Voltage vs. Capacity (1st Cycle)

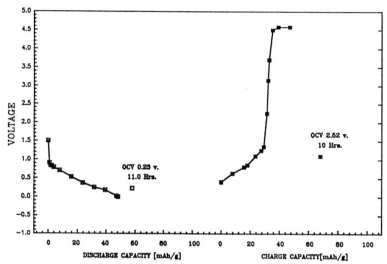

Figure 3: PET Coke - Voltage vs. Capacity (2nd Cycle)

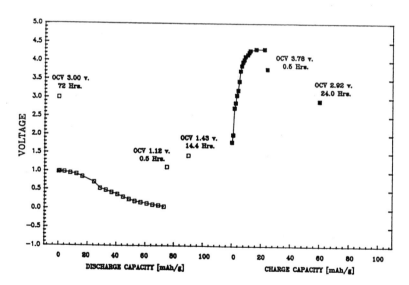

Figure 4: Partially Graphitized Coke
Voltage vs. Capacity (1st Cycle)

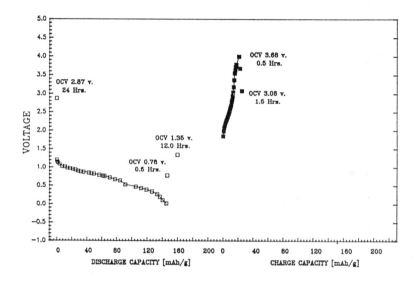

Figure 5: Fully Graphitized Coke
Voltage vs. Capacity (1st Cycle)

TABLE 3

Pan and Pitch Fiber Capacity at 1 mA/cm² in 1M LiAsF₆ in 50/50 Volume % PC/DME

Carbon Type	Weight (g)	Discharge Hrs to OV	Charge Hrs to 1.5V	Discharge Capacity (mAh/g)	Charge Capacity (mAh/g)	Discharge Formula	Charge Formula
AMOCO PAN T-50	0.136	1.0	0.85	29	25	$Li_{0.08}C_6$	$Li_{0.07}C_6$
Grafil Pan XAS	0.168	5.2	3.20	124	76	$Li_{0.33}C_6$	$Li_{0.20}C_6$
Grafil PAN HMS	0.216	6.3	5.70	117	106	$Li_{0.31}C_6$	$Li_{0.28}C_6$
EG+G PAN Foam	0.112	11.7	1.10	418	39	$Li_{1.12}C_6$	$Li_{0.10}C_6$
Mesophase Pitch P-55	0.104	1.5	1.30	58	50	$Li_{0.16}C_6$	$Li_{0.13}C_6$
Mesophase Pitch P-100	0.096	0.9	0.00	38	0	$Li_{0.10}C_6$	Li_0C_6

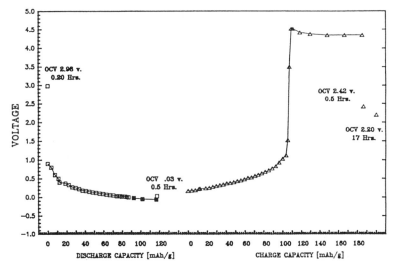

Figure 6:   HM-S PAN Fiber
Voltage vs. Capacity   (1st Cycle)

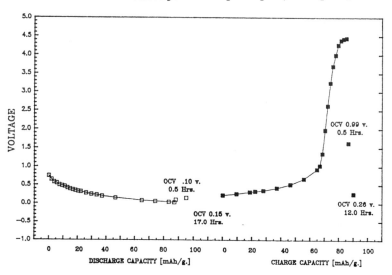

Figure 7:   HM-S PAN Fiber
Voltage vs. Capacity   (3rd Cycle)

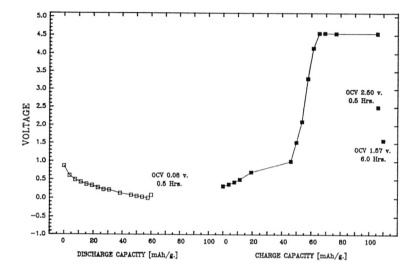

Figure 8:   P-55 Pitch Fiber
            Voltage vs. Capacity   (1st Cycle)

LAYERED-STRUCTURE BC_2N AS A NEGATIVE ELECTRODE MATRIX FOR RECHARGEABLE LITHIUM BATTERIES

Masayuki MORITA, Tomoyuki HANADA, Hiromori Tsutsumi, and Yoshiharu MATSUDA

Department of Applied Chemistry and Chemical Engineering, Faculty of Engineering, Yamaguchi University, Tokiwadai, Ube 755, Japan

Masayuki KAWAGUCHI

Ube Research Center, Central Glass Co. Ltd., Okiube, Ube 755, Japan

ABSTRACT

A new compound with graphite-like structure, BC_2N, has been synthesized by a vapor phase reaction of acetonitrile (CH_3CN) and borontrichloride (BCl_3). The electrochemical behavior of BC_2N was investigated in propylene carbonate (PC)-based solutions containing lithium (Li) salts. Cathodic current increased and corresponding anodic current peaks were observed in the cyclic voltammetric curves of the BC_2N electrode, which showed the electrochemical intercalation/ deintercalation of Li from/to the electrolytes, respectively. The polarization behavior of Li with BC_2N matrix depended on the solvent as well as the salt of the electrolytic solution. The coulombic efficiency for charge/discharge cycling under a constant current was over 90 % in $LiClO_4$/PC. The results indicate that BC_2N is suitable as a negative electrode matrix for rechargeable Li batteries.

INTRODUCTION

A key technology for realizing secondary lithium (Li) batteries working at ambient temperature with high electric capacities is to ensure high rechargeability of Li negative electrode in nonaqueous electrolytes. The use of metallic Li as the negative electrode often causes some problems including the formation of Li dendrite and the reaction of deposited Li with the electrolyte, which lead to degradation of the rechargeability of the battery (1). The use of Li-alloys such as Li-Al, Li-Wood's metal, etc. is efficient for improving the cycleability of the electrode (2), but there still remains such problems as lower capacity and energy density for the batteries with alloy negatives.

Some carbon materials can take alkali metals in their layered graphite structures to form graphite-intercalation-compounds (GICs). These structural characteristics provide a possibility of carbon materials being utilized as the electrode matrix of the rechargeable battery. Also light weights of carbon materials would be advantageous to the electrode matrices of high energy-density batteries. Thus much attention has been directed to the electrochemical intercalation/deintercalation processes of carbon materials in organic media (3-7), and then some prototypes of rechargeable Li batteries with carbon electrodes have been demonstrated (8,9).

Besides carbon materials, other layered- or tunnel structure compounds with small molecular weights are usuable for this purpose. We have synthesized a new layered structure compound, BC_2N, by a chemical vapor deposition (CVD) method, and attempted to apply it to the negative electrode matrix of a rechargeable Li battery. In this paper, fundamental charge/discharge behavior of the negative electrode with a BC_2N matrix was investigated in propylene carbonate-based organic electrolytes.

EXPERIMENTAL

The preparation of BC_2N was based on the method proposed by Bartlett, et al. (10). Acetonitrile (CH_3CN) and borontrichloride (BCl_3) were reacted in a vapor phase at 1000°C (Eq. [1]).

$$CH_3CN + BCl_3 \rightarrow BC_2N + 3HCl \qquad [1]$$

Dark brown or black powder was deposited in a tubular reactor. The deposite was further ground to fine powder and then heated at 1000°C in dry N_2 for 1 h. The resulting BC_2N was characterized by elementl analysis and X-ray diffractometry (XRD).

The electrochemical properties of BC_2N was investigated

102

in organic electrolyte solutions.   The BC_2N working electrode used generally was prepared by mixing the BC_2N powder (85 mg) with poly(tetrafuluoroethylene) (PTFE, 15 mg) as a binder, followed by molding under a pressure of 2.2 x 10^3 kg cm^{-2} to form a tablet (13 mmϕ, 0.5 mm thick).   An undivided Pyrex-glass beaker cell (inner volume: 50 cm^3) was used as the test cell.   A large surface area Li sheet and a Li chip were used as the counter and the reference electrodes, respectively.   Another test cell was bobbin-type cell, in which the working electrode was a 16 mmϕ disk consisting of 51.5 mg BC_2N and 9.0 mg PTFE, and the counter electrode was a Li disk.   The solvents of the electrolyte solutions were propylene carbonate (PC) and a 1:1 (by vol.) mixture of PC and 1,2-dimethoxyethane (DME), which were used as received (Mitsubishi Petrochemical, Battery Grade,). Vacuum-dried $LiBF_4$, $LiPF_6$ (Morita Chemical Industries) and $LiClO_4$ (Ishizu Pharmaceutical) were mainly used as the electrolytic salts.   In some experiments, $NaClO_4$ and $(C_2H_5)_4NBF_4$ were used for comparison.   Cyclic voltammetry (CV) and constant-current charge/discharge cycling were carried out under a dry argon atmosphere at room temperature (18-24℃).

The Li content in the electrode after cathodic polarization was determined by a quantitative analysis using atomic absorption spectrometry (AAS).   The BC_2N electrode was treated with concentrated hydrochloric acid to elute Li incorporated in the electrode during the polarization, and then the eluate was analyzed by AAS.

RESULTS AND DISCUSSION

The chemical composition of the product obtained by the CVD reaction of CH_3CN and BCl_3 (Eq. [1]) was almost consistent with stoichiometric BC_2N.   The XRD pattern of the BC_2N powder exhibited broad peaks which are typical of small particle graphite with partly disordered structure. It also showed that the interlayer spacing, $c_0/2$, of BC_2N was 0.354 nm, which was slightly larger than that of natural graphite (0.335 nm).   Though the real atomic configuration in BC_2N have not been clear yet, a possible arrangement of atoms in a basal plane is schematically represented as Fig.1.   The resistivity of the BC_2N tablet pressed with PTFE binder was 10^2 - 10^3 Ω cm.

Figure 2 shows CV curves (scan rate: 1 mV s^{-1}) in PC dissolving 1 mol dm^{-3} of Li salts.   Increases in the cathodic currents at 1.0 V (vs. Li/Li$^+$) or below are mainly based on the electrochemical intercalation (or doping) of Li into the BC$_2$N matrix. The anodic currents with broad peaks at ca. 1.0 V are accompanied with deintercalation (or undoping) of Li from the matrix. These could be formulated as Eq. [2],

$$(BC_2N)_n + Li^+ + e^- \rightleftharpoons Li \cdot (BC_2N)_n \qquad [2]$$

where n denotes the number of the BC$_2$N unit. The rates of intercalation/deintercalation processes depended on the electrolytic salt.

Figure 3 shows CV curves in the solutions consisting of a mixed PC+DME (1:1 by vol.) solvent. Voltammetric currents in mixed PC+DME solutions were higher than those in simple PC systems, and the difference in the shape of the CV curve among the electrolytic salts were somewhat less than that observed in the solutions using pure PC solvent. The solvent blending effects on the polarization behavior are often detected not only for simple deposition/dissolution reaction of a metallic Li electrode (11,12) but also for the electrodes with topochemical reactions (7,13). The structure and properties of the solutions would influence the reaction rate of Eq. [2] though details of the solvent blending effects are not clear. On the other hand, CVs in the solutions containing NaClO$_4$ and $(C_2H_5)_4$NBF$_4$ salts had no visible current peaks. This suggests that the size of cation is of great importance for the electrochemical intercalation/deintercalation processes of BC$_2$N.

Figure 4 shows typical charge/discharge curves of BC$_2$N measured in a glass beaker cell under a galvanostatic condition in PC and mixed PC+DME electrolytes. The mass of BC$_2$N in the electrode was 85 mg and the cycling current was 1.0 mA cm^{-2}. The charging process corresponds to cathodic intercalation of Li into the BC$_2$N matrix, and the discharging is the reverse reaction. The electrolytic solutions in which higher current densities were observed in CV seems to be preferable to the charge/discharge cycling of Li on BC$_2$N. It is notable that variations in the charge/discharge curves with repeated cycles were only slight. This would be a special feature of the BC$_2$N matrix.
The coulombic efficiency for cycling can be defined as

$$E_{ff} = (Q_{dis}/Q_{ch}) \times 100 \ [\%] \qquad\qquad [3]$$

where Q_{dis} and Q_{ch} are the quantities of electricity passed for discharging and charging, respectively.   In this case, Q_{ch} was kept constant (1.8 C cm^{-2}) through the cycles. Variations of the efficiency with cycling are shown in Fig. 5.   The efficiency in LiClO$_4$/PC was 90 % or higher under this cycling condition and did not show any degradation after the repeated cycles.   However, the efficiency in LiBF$_4$/PC was very low.   This is probably due to some side reaction (eg., cathodic decomposition of the electrolyte on the BC$_2$N matrix).   On the other hand, cycling in the mixed PC+DME solutions gave stable charges in the coulombic efficiency.   The difference in the efficiency among the electrolytic salts was small for the cycling in the mixed PC+DME solutions.

The depth of charge, or doping level, in Fig. 4 is about 65% in Eq. [2].   Furthermore, higher doping level experiments were carried out by using a bobbin-type cell consisting of the BC$_2$N working and a Li counter electrodes (BC$_2$N/Li).   Here the mass of BC$_2$N was 51.5 mg and the current density was 25 μA cm^{-2}.   That is, the cycling condition in this case was much milder than that shown in Fig. 4.   For this bobbin-type cell, a deeper charging level, ie. a longer charging time of 330 h, was achieved, which gives a theoretical doping level of n = 1.71 for Li·(BC$_2$N)$_n$. This was approximately equivalent to the value of a stoichiometric GIC (C$_6$Li).

The amounts of Li incorporated in BC$_2$N were analyzed for the electrodes after charging under a constant current of 1.0 mA cm^{-2} with different quantities of electricity. Figure 6 shows the relation between the amount of Li incorporated and the quantity of electricity passed.   The plots almost lie on the theoretical line except for larger quantities of electricity.   Such a linear relation proves that equivalent amounts of Li to the charged electricity were quantitatively incorporated in BC$_2$N (ie., with ~100 % current efficiency for charging).

Figure 7 is a series of XRD patterns for the BC$_2$N electrode before and after cathodic charging (electro-chemical intercalation) in LiClO$_4$/PC+DME.   The BC$_2$N electrode before use has a sharp peak ($2\theta \simeq 17.8°$) and two broad peaks (2θ = 25.1° and 42 - 44°).   The former sharp peak was based on the PTFE binder in the electrode.   The

105

latter broad peaks were assigned to the diffraction from (002) and (10) planes in the layered structure of BC_2N. The XRD patterns after the cathodic charge showed another sharp peak. This was assigned to Li compounds (Li_2CO_3) which might be reaction products of deposited Li with the electrolyte solution and CO_2 in air during the XRD measurements. Besides these, the broad peak based on the (002) diffraction shifted apparently to lower angles. The degree of angle shift became more significant in higher doping levels (or larger amounts of electricity for charging).

The apparent interplanar distance, $\underline{d}$, was calculated from the peak angle, and its variation with the charged electricity is summarized in Table 1. The apparent d value tended to increase with charging. As the amount of Li in BC_2N was proportional to the charged electricity (Fig. 6), it is obvious that the structural strain by the cathodic charge was caused by the electrochemical intercalation of Li into the BC_2N matrix. However, the degree of increases in the $\underline{d}$ value was not so high. Thus, relatively small changes in XRD profiles by the cathodic charge would be related to the cycling result shown in Fig. 4 that only slight changes were observed in the charge/discharge curves even after the repeated cycles. It can be concluded that the present BC_2N is a possible material of the electrode matrix for rechargeable Li batteries. Details of the electrochemical properties of this material are now under investigation.

ACKNOWLEDGMENT

The authors are deeply grateful to Prof. N. Bartlett, Department of Chemistry, University of California, Berkeley, for his great interest and discussion.

REFERENCES

1. M. Morita and Y. Matsuda, in "Practical Lithium Batteries", Y. Matsuda and C. R. Schlaikjer, Editors, Chap. 19, p. 87, JEC Press Inc., Cleveland (1988).
2. Z. Takehara, Progress in Batteries & Solar Cells, 7, 252 (1988).
3. R. Kanno, Y. Takeda, T. Ichikawa, K. Nakanishi, and O. Yamamoto, J. Power Sources, 26, 535 (1989).

4. M. Mohri, N. Yanagisawa, Y. Tajima, H. Tanaka,
   T. Mitate, S. Nakajima, M. Yoshida. Y. Yoshimoto,
   T. Suzuki, and H. Wada, ibid., 26, 545 (1989).
5. K. Sawai, T. Ohzuku, and T. Hirai, Chem. Express, 5,
   387 (1990).
6. R. Fong, U. Sacken, and J. R. Dahn, J. Electrochem.
   Soc., 137, 2009 (1990).
7. M. Morita, N. Nishimura, H. Tsutsumi, and Y. Matsuda,
   Chem. Express, 6, 619 (1991).
8. K. Inada, K. Ikeda, S. Inomata, T. Nishii,
   M. Miyabayashi, and H. Yui, in "Practical Lithium
   Batteries", Y. Matsuda and C. R. Schlaikjer, Editors,
   Chap. 21, p. 96, JEC Press Inc., Cleveland (1988).
9. H. Imoto, H. Azuma, A. Omaru, and Y. Nishi, Extended
   Abstracts of the 58th Annual Meeting of the
   Electrochemical Society of Japan (2F11), p. 158 (1991).
10. J. Kouvetakis, T. Sasaki, C. Shen, R. Hagiwara,
    M. Lerner, K. M. Krishnan, and N. Bartlett, Synth.
    Metals, 34, 1 (1989).
11. S. Tobishima and T. Okada, Electrochim. Acta, 30,
    1715 (1985).
12. M. Morita and Y. Matsuda, J. Power Sources, 20,
    299 (1987).
13. M. Uchiyama, S. Slane, E. Plichta, and M. Salomon,
    ibid., 20, 279 (1987).

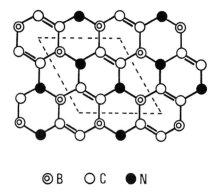

Fig. 1 A possible configuration of atoms in the basal plane of BC_2N.

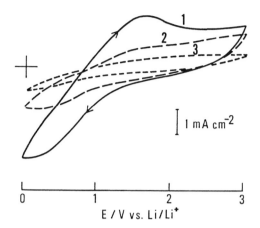

Fig. 2 Cyclic voltammograms of BC$_2$N electrode in PC
containing 1 mol dm^{-3} Li salts,
1: LiClO$_4$, 2: LiPF$_6$, 3: LiBF$_4$,
Scan rate: 1 mV s^{-1}, The amount of BC$_2$N: 85 mg,
Each curve indicates the voltammogram after 9 cycles

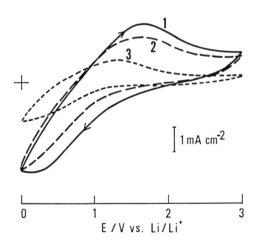

Fig. 3 Cyclic voltammograms of BC$_2$N electrode in mixed
PC+DME(1:1 by vol.) containing 1 mol dm^{-3} Li salts,
1: LiClO$_4$, 2: LiPF$_6$, 3: LiBF$_4$,
Scan rate: 1 mV s^{-1}, The amount of BC$_2$N: 85 mg,
Each curve indicates the voltammogram after 9 cycles

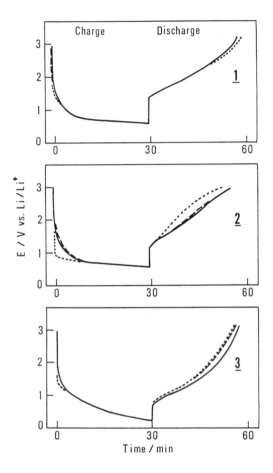

Fig. 4 Charge/discharge curves of BC_2N electrode in PC and
PC+DME(1:1 by vol.) containing 1 mol dm^{-3} Li salts,
1: $LiClO_4$/PC, 2: $LiClO_4$/PC+DME, 3: $LiPF_6$/PC+DME,
$i_{ch} = i_{dis} = 1.0$ mA cm^{-2}, $Q_{ch} = 1.8$ C cm^{-2},
------: 1st cycle, —— : 5th cycle, —— : 10th cycle,
The amount of BC_2N: 85 mg, Current: 1.0 mA cm^{-2}.

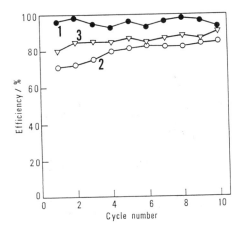

Fig. 5 Coulombic efficiencies for charge/discharge cycling
of BC_2N electrode in PC and PC+DME(1:1 by vol.)
containing 1 mol dm^{-3} Li salts,
1: $LiClO_4$/PC, 2: $LiClO_4$/PC+DME, 3: $LiPF_6$/PC+DME,
$i_{ch} = i_{dis} = 1.0$ mA cm^{-2}, $Q_{ch} = 1.8$ C cm^{-2},
The amount of BC_2N: 85 mg.

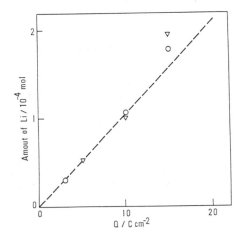

Fig. 6 A relation between the amount of Li incorporated in
BC_2N and the quantity of charged electricity,
$\bigcirc$: $LiClO_4$/PC, $\bigtriangledown$: $LiClO_4$/PC+DME,
Broken line: theoretical relation,
The amount of BC_2N: 85 mg, Current: 1.0 mA cm^{-2}.

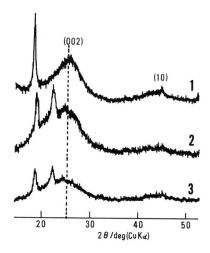

Fig. 7 XRD patterns of BC_2N electrode before and after cathodic charge at 1.0 mA cm^{-2} in $LiClO_4$/PC+DME, 1: Before charge, 2: After 5.3 C cm^{-2} charge, 3: After 21.1 C cm^{-2} charge, Charging current: 1.0 mA cm^{-2}.

Table 1 XRD data for BC_2N before and after cathodic charge[a]

Amount of electricity/C cm^{-2}	2θ /deg	d/nm
0	25.1	0.354
5.3	23.7	0.375
21.1	23.2	0.383

a) in $LiClO_4$(1 mol dm^{-3})/PC+DME(1:1 by vol.), Current: 1.0 mA cm^{-2}.

LITHIUM METAL-FREE RECHARGEABLE $Li_{1+x}Mn_2O_4$ / CARBON CELLS

D. Guyomard and J.M. Tarascon

Bellcore, Red Bank, NJ 07701

ABSTRACT

Ambient temperature secondary lithium cells have safety problems due to the use of highly reactive lithium metal. To alleviate this problem, the concept of a "rocking-chair battery", that uses two insertion electrodes, has been proposed. The behavior of such a type of cell, using the spinel $LiMn_2O_4$ as the positive electrode and a disordered carbon, petroleum coke, as the negative, is presented. The $LiMn_2O_4/C$ batteries show high specific energies (3 times and 1.5 times higher than that of Ni-Cd and Ni-metal hydride batteries, respectively), high power rate capabilities (comparable to Ni-Cd batteries) and promising cycle life even at 55°C. The large irreversible capacity loss at first cycle, common to all rocking-chair batteries based on carbon as the negative electrode, can in the present case be overcome by using as the positive electrode the composition $Li_{1+x}Mn_2O_4$, where x acts as a Li reservoir. We show that rocking-chair cells using a positive electrode of composition $x \cong 0.1$ deliver about 10% higher specific capacity and specific energy than cells using a positive electrode of composition $x = 0$.

INTRODUCTION

Among the rechargeable batteries, secondary lithium cells using lithium, the most electropositive and the lightest metal, as the negative electrode can theoretically deliver the highest specific energy. Room temperature secondary lithium cells have been studied for more than 15 years (1-3) and despite their promises are still not commercially available because of several problems. The most important one being a safety problem associated to the use of pure Li metal and more specifically to the dendritic regrowth of Li upon cycling that might

113

short-circuit the cell (4).

A more advanced and inherently safe approach to lithium batteries consists in replacing lithium metal with a lithium intercalation compound, leading to Li metal-free rechargeable batteries, the so-called "rocking-chair" batteries. This approach, proposed a decade ago (5-9), was never successfully realized until the recent announcement by two battery companies (10,11) of the future commercialization of the "rocking-chair" technology. The rocking-chair cells already produced by Sony Energytec and announced by Moli Energy are both using carbon as the negative electrode and $LiCoO_2$ and $LiNiO_2$ as the positive electrode, respectively.

We recently proposed another rocking-chair system (12) based on the spinel lithium manganese oxide for the positive and a disordered carbon, petroleum coke, for the negative. Our system shows similar performances to Sony' and Moli' systems. However, the use of the spinel manganese oxide material offers the following advantages: 1) a lower overall electrode cost coming from an easier synthesis (only 1 step, lower temperature) than that of $LiNiO_2$, and the use of Mn which is more abundant and cheaper than Co, and 2), but not the least important, is that manganese-based oxide materials are extensively used in primary cells, so that toxicity (if any) or recycling problems are already well known.

In this paper, we review the behavior of the $LiMn_2O_4/C$ cells and show how the second lithiation plateau of the spinel can be used to further increase the specific capacity and energy of the manganese-based rocking-chair batteries.

EXPERIMENTAL

$LiMn_2O_4$ materials were prepared in air by reacting stoichiometric amounts of Li_2CO_3 and MnO_2 powders. Three consecutive anneals at 800°C (24 hours each) lead to $LiMn_2O_4$ powders of 1 to 2 μm particle size. Petroleum coke (200/400 mesh, i.e particle size between 38μm and 75μm) was obtained from Conoco. Composite porous electrodes of the following massic compositions, 89% $LiMn_2O_4$, 10% black carbon and 1% EPDM binder for the positive electrode, and 94% coke, 5% black carbon and 1% EPDM binder for the negative electrode, were found to give the best electrochemical results in term of high capacity, high reversibility and low polarization when cycled separately versus Li metal.

The electrolyte used in this study was made by adding $LiClO_4$ salt, in a 1 molar concentration (1M), to a solvent consisting of an equal mixture of ethylene carbonate (EC) and diethoxyethane (DEE). The solvents and the salt have been dried using conventional procedures to prevent any water contamination. The water content of the electrolyte was as low as 40 ppm and its ionic conductivity was about 10 mS/cm at room temperature, as determined by means of a Karl-Fisher coulometer and a high frequency impedance bridge, respectively.

Standard swagelock test cells and three-electrode test cells were used to study the cycling behavior of either the Li or the rocking-chair batteries presented herein. These test cells were assembled in a helium glove box by sandwiching a glass paper soaked in the electrolyte between the positive and the negative electrodes.

RESULTS AND DISCUSSION

A rocking-chair battery uses intercalation compounds for the positive and negative electrodes. Thus, a main requirement prior to the assembly of such a cell is to study and to optimize the electrochemical behavior of each composite intercalation electrode versus lithium.

1- Behavior of each electrode versus Li

The cycling behavior, at room temperature between 3.5 V and 4.5 V, and at 55°C between 3.5 V and 4.3 V, of an optimized $LiMn_2O_4/EC$ + DEE (50:50) + 1M $LiClO_4$/Li cell is shown in Figure 1. These results are in good agreement with previous work (13-15). A lower charge cut-off voltage was used at high temperature in order to avoid any oxidation of the electrolyte. Note that independent of the operating temperature, 0.8 Li can be removed reversibly from the spinel structure at an average voltage of 4.1 V versus Li.

Figure 2 shows the cycling behavior of an optimized C/EC + DEE (50:50) + 1M $LiClO_4$/Li at the same two operating temperatures. When discharged down to 0.02 V versus Li, petroleum coke intercalates reversibly, at an average potential of 0.4 V versus Li, 0.083 Li per carbon atom, leading to the maximum composition $Li_{0.5}C_6$. The main feature of this data is the large irreversible capacity loss ($\cong 23\%$ at 25°C) at first discharge. This capacity loss is related to some superficial reaction with the solvent, leading to a SEI (solid electrolyte interface) layer on the carbon surface (16). It is this large capacity loss at first cycle that reduces the whole capacity of rocking-chair cells based on carbon as the negative electrode.

115

2- Behavior of the $LiMn_2O_4/C$ rocking-chair cell

Any rocking-chair cells can be viewed as consisting of two lithium sponges with, in the present case, $LiMn_2O_4$, the high voltage compound, acting as the positive electrode and carbon, the low voltage compound, acting as the negative electrode. When such a cell operates, the Li ions are shuttled back and forth from one sponge to the next and optimization is reached when the mass of the two sponges is adjusted, so that one is fully lithiated when the other one is fully delithiated. To determine this optimum condition we used a three-electrode cell (lithium being the third electrode), that allows to follow the intrinsic behavior of each electrode material (its voltage versus Li), while monitoring the output voltage of the rocking-chair battery (Figure 3a). In addition, the three-electrode cell turns out to be a powerful tool to understand the consequence of an overcharge or an overdischarge and to adjust the charge and discharge cut-off voltages of the cell.

The typical voltage-capacity curves obtained for an optimized cell (i.e. a cell with the mass of the positive material being 2.1 times the mass of the negative material), are shown in Figure 3b. At the end of the first charge, the positive electrode is fully delithiated while the negative electrode is lithiated with a voltage close to 0.02 V versus Li, showing that the masses of the two electrodes balance perfectly. At the end of the first discharge, as shown by the voltage of each electrode versus Li, carbon, is limiting the capacity of the whole cell, as expected because of the 23% capacity loss previously mentioned.

Figure 4 shows the scenario for an overcharged rocking-chair cell. The plateau at 4.55 V versus Li, corresponding to irreversible capacity (17), indicates the oxidative decomposition of the electrolyte occurring at the $LiMn_2O_4$ electrode. Simultaneously, at the negative, more Li is intercalated in petroleum coke (up to a composition $Li_{0.7}C_6$) and then, Li plating on carbon occurs, as can be seen by the voltage plateau close to 0 V versus Li and separately confirmed by measuring, using in-situ x-ray diffraction, the variation of the interlayer distance of carbon during the cell discharge down to very low voltage (17). In short, an overcharge of the battery results in an increase of the internal pressure of the cell because of gaseous oxidative products of the solvent and in the formation of dispersed metallic lithium on the carbon surface with the overall result being the rapid loss of cycling reversibility. However the control of the charge cut-off voltage at values lower than 4.4 V, allows the cell to operate **in completely safe conditions**.

The cycling behavior at room temperature of a $LiMn_2O_4/C$ cell using optimized conditions is presented in Figure 5. The cell delivers its capacity between 4.35 V and 2.2 V, at an average voltage of 3.7 V. The capacity, calculated at the second discharge, is 67 mAh per gram of both active electrode masses.

116

These values lead to energy densities of 250 mWh per gram of active electrodes or 230 mWh per gram of composite electrodes. The performances of these batteries are compared to that of parent systems and competitive aqueous systems in Table I. The main point is that rocking-chair cells using a lithiated oxide and a carbon electrode have roughly the same specific capacity as Ni-Cd cells with an output voltage three times higher, so they deliver energy densities about 3 times higher than that of Ni-Cd and 1.5 times higher than that reported for Ni-metal hydride cells. Moreover, when using thin electrodes (< 100 μm), high specific power can be achieved (17), of the same order of magnitude of that delivered by Ni-Cd (more than 100 mW/g with a specific energy more than half of the maximum one).

The capacity changes for a $LiMn_2O_4/C$ cell as a function of the cycle number is shown in Figure 6 for different operating temperatures. The capacity always decreases during the 10 to 20 first cycles by a factor of 11% to 20%, and then stabilizes. The capacity measured at 55°C is smaller than at 25°C because of the lower charge cut-off voltage (4.3 V instead of 4.5 V), used in order to maintain the voltage of $LiMn_2O_4$ far from electrolyte oxidation. Finally, note that when the cell is cycled at -20°C at a rate of C/5, its capacity is still 85% of the room temperature capacity.

3- Interest of a Li reservoir

We previously showed that it is the carbon electrode, because of the large irreversible capacity loss occurring at first cycle, that limits the capacity of the $LiMn_2O_4/C$ rocking-chair cell. This is a common problem to all the rocking-chair batteries using carbon as the negative electrode, i.e. $LiNiO_2/C$ and $LiCoO_2/C$ cells.

$LiMn_2O_4$ and $LiNiO_2$ can intercalate a second lithium per formula unit at constant voltages of 3 V and 1.9 V versus Li, respectively, leading to single phase materials of composition $Li_2Mn_2O_4$ and Li_2NiO_2. These rich lithium phases, if non moisture sensitive, could be used as a Li reservoir to overcome the irreversibility inherent to carbon at first cycle. Because of its voltage higher than 2.5 V versus Li (the voltage of reduction of water), $Li_{1+x}Mn_2O_4$ should be non-moisture sensitive and thereby easily handable in air, in contrast to the $Li_{1+x}NiO_2$ that forms at a voltage lower than 2.5 V versus Li.

When prepared by using n-butyl lithium, $Li_2Mn_2O_4$ was reported to be highly hygroscopic (21), probably due to some moisture sensitive decomposition products resulting from the use of a too strong reducing agent. We recently reported on the use of a mild reducing agent, LiI, that allows to prepare non-moisture sensitive $Li_2Mn_2O_4$ (12). Figure 7 shows the voltage of the $LiMn_2O_4$

117

to $Li_2Mn_2O_4$ transformation and the calculated voltage of a I_3^-/I^- solution as a function of the state of the redox reaction between $LiMn_2O_4$ and LiI at 82°C. From this plot, one can easily find the experimental conditions to prepare either $Li_2Mn_2O_4$ or $Li_{1+x}Mn_2O_4$ of any wanted composition x. The validity of the model has been checked experimentally. All the experimental procedure can be done in air (i.e., the handling of LiI, of acetonitrile, the filtration of resulting powders), as well as the storage of the resulting $Li_{1+x}Mn_2O_4$ powders, with no influence on the material electrochemical behavior.

Next, we show that the Li in excess (x) with respect to the nominal composition $LiMn_2O_4$ in the newly prepared $Li_{1+x}Mn_2O_4$ material can act as a lithium reservoir so the performances of our lithium manganese oxide / carbon cells can be further improved. For this to happen, the value of x has to be adjusted, so that the capacity loss during the first cycle of the positive electrode (starting with the composition $Li_{1+x}Mn_2O_4$ and ending with the composition $LiMn_2O_4$) is the same as that occurring on carbon at first cycle. This condition was achieved when using the initial composition $Li_{1.1}Mn_2O_4$. Then the three-electrode set-up has been used to further optimize the cell. The voltage versus capacity traces for an optimized $Li_{1.1}Mn_2O_4/C$ cell are shown in Figure 8. The voltage of both electrodes versus Li at the end of the first charge indicates that the electrode masses are well balanced while the voltage of both electrodes versus Li at the end of the first discharge shows that the whole reversible capacity of both electrodes is fully used after the first charge. For a well defined amount of petroleum-coke, because of the higher specific capacity of $Li_{1+x}Mn_2O_4$ compared to $LiMn_2O_4$, a lower mass of $Li_{1+x}Mn_2O_4$ than $LiMn_2O_4$ is needed to balance the rocking-chair cell. It results that rocking-chair cells based on $Li_{1.1}Mn_2O_4$ instead of $LiMn_2O_4$ as the positive show a 9% increase of both specific capacity and specific energy.

Finally, when $Li_2Mn_2O_4$ is used instead of $LiMn_2O_4$ in a carbon rocking-chair cell, an increase (calculated from the first discharge, Figure 9) of the specific capacity and specific energy by 53% and 20%, respectively, is observed. This increase results from the higher capacity of the positive electrode $Li_2Mn_2O_4$ compared to $LiMn_2O_4$. It is worth to mention that the use of the lower voltage plateau of the spinel is accompanied by a sharp and large drop (1 V) in the voltage composition curves, that makes these cells unsuitable for some applications in contrast to the $LiMn_2O_4/C$ ones.

In summary we have reviewed the electrochemical behavior of the $LiMn_2O_4/C$ rocking-chair cells that allow for specific energy 3 times higher than Ni-Cd cells and have shown how to exploit the presence of the lithium-rich spinel phase ($Li_{1+x}Mn_2O_4$) to improve their performance further. Note that the same cannot be done with $LiNiO_2$ or $LiCoO_2$ materials because $Li_{1+x}NiO_2$ is water-sensitive and consequently cannot be handled in air, and $Li_{1+x}CoO_2$ compositions do not exist. At the time of this study the electrolyte consisting of 1M $LiClO_4$ in EC + DEE (50:50) was giving the best results. Since, another electrolyte that does not contain a lithium perchlorate salt and that allows us to operate in a safer mode at high voltages, has been identified and will be reported elsewhere (22). Finally, it is important to recognize that the performances of the present rocking-chair cells ($LiCoO_2/C$, $LiNiO_2/C$ and $LiMn_2O_4/C$) could be highly improved if one could replace the negative carbon electrode by another intercalation compound with higher specific capacity and able to reversibly intercalate lithium at low voltages. We are presently investigating this new avenue of research.

ACKNOWLEDGEMENTS

We would like to thank J.H. Wernick, W.R. Mc Kinnon, F.K Shokoohi, P. Warren, G. Baker and S. Colson for helpful discussions and J. Gural for the technological constructions needed in this work. We also thank Conoco Inc. for the petroleum coke.

REFERENCES

1. G. Eichinger and J.O. Besenhard, *J. Electroanal. Chem.*, 72, 1 (1976).
2. D.W. Murphy and P.A. Christian, *Science*, 205, 651 (1979).
3. J.P. Gabano, "Lithium Batteries", Academic Press Ltd, London (1983).
4. D.P. Wilkinson, J.R. Dahn, U. Von Sacken and D.T. Fouchard, papers 53 and 54, 178th Meeting of the Electrochemical Society, Seattle, Washington, 1990.
5. D.W. Murphy, F.J. DiSalvo, J.N. Carides and J.V. Waszczak, *Mat. Res. Bull.*, 13, 1395 (1978).
6. D.W. Murphy and J.N. Carides, *J. Electrochem. Soc.*, 126, 349 (1979).
7. M. Lazzari and B. Scrosati, *J. Electrochem. Soc.*, 127, 773 (1980).
8. M. Armand, in "Materials for Advanced Batteries", D.W. Murphy, J. Broadhead and B.C.H. Steele (Eds), p. 145, Plenum Press, New York (1980).
9. K. Mizushima, P.C. Jones, P.J. Wiseman, and J.B. Goodenough, *Mat. Res. Bull.*, 15, 783 (1980).
10. T. Nagaura, 4th International Rechargeable Battery Seminar, Deerfield Beach, Florida, 1990.
11. J.R. Dahn, U. Von Sacken and R. Fong, paper 42, 178th Meeting of the Electrochemical Society, Seattle, Washington, 1990.
12. J.M. Tarascon and D. Guyomard, *J. Electrochem. Soc.*, 138, 2864 (1991).
13. M.H. Rossouw, A. de Kock, L.A. de Picciotto, M.M. Tackeray, W.I.F. David and R.M. Ibberson, *Mat. Res. Bull.*, 25, 173 (1990).

14. T. Ohzuku, M. Kitagawa and T. Hirai, *J. Electrochem. Soc.*, 137, 769 (1990).
15. J.M. Tarascon, E. Wang, F.K. Shokoohi, W.R. McKinnon and S. Colson, *J. Electrochem. Soc.*, 138, 2859 (1991).
16. R. Fong, U. Von Sacken and J.R. Dahn, *J. Electrochem. Soc.*, 137, 2009 (1990).
17. D. Guyomard and J.M. Tarascon, *J. Electrochem. Soc.*, to be published in April 1992.
18. T. Sakai, H. Miyamura, N. Kuriyama and H. Ishikawa, paper 105, 180th Meeting of the Electrochemical Society, Phoenix, Arizona, 1991.
19. J.R. Dahn, U. Von Sacken, M.W. Juskow and H. Al-Jabany, *J. Electrochem. Soc.*, 138, 2207 (1991).
20. E. Plichta, M. Salomon, S. Slane, M. Uchiyama, D. Chua, W.B. Ebner and H.W. Lin, *J. Power Sources*, 21, 25 (1987).
21. A. Mosbah, A. Verbaere and M. Tournoux, *Mater. Res. Bull.*, 18, 1375 (1983).
22. D. Guyomard and J.M. Tarascon, *J. Electrochem. Soc.*, to be submitted.

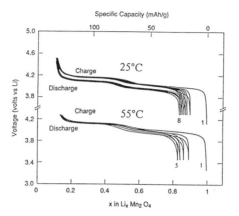

Figure 1: Typical cycling behavior at 25°C (a) and at 55°C (b) of $LiMn_2O_4/EC + DEE(50:50) + 1MLiClO_4/Li$ cells at a rate of C/10.

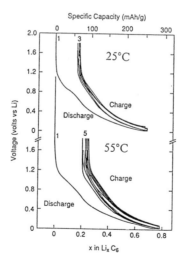

Figure 2: Typical cycling behavior at 25°C (a) and at 55°C (b) of C (petroleum coke)/$EC + DEE(50:50) + 1MLiClO_4/Li$ cells at a rate of C/20.

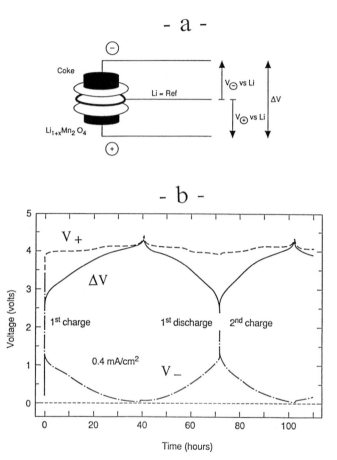

Figure 3: The schematic of a three-electrode cell is shown in (a) and the first cycles at room temperature of a $LiMn_2O_4/EC + DEE(50:50) + 1MLiClO_4/C$ (petroleum coke) three-electrode cell, under constant current, are shown in (b).

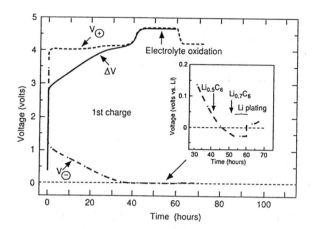

Figure 4: Behavior of a $LiMn_2O_4/EC + DEE(50:50) + 1MLiClO_4/C$ rocking-chair cell at room temperature when overcharged to a voltage higher than 4.55V, as measured with the three-electrode experiment. The insert shows an expand of the carbon electrode voltage vs. Li during the overcharge of the cell.

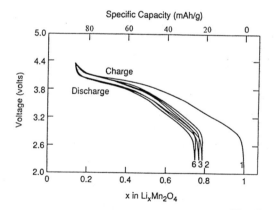

Figure 5: Cycling behavior at room temperature of a $LiMn_2O_4/EC + DEE(50:50) + 1MLiClO_4/C$ rocking-chair cell a rate of C/10 with the mass of the positive electrode being 2.1 times that of the negative electrode.

Table I

The performance of different types of aqueous and Li rechargeable batteries. For the Ni- and Mn-based rocking-chair cells, since no cylindrical cells have been made yet, a realization factor of 33% (20) on the specific capacity and energy, has been assumed for the extrapolation.

	Capacity	Average Voltage	Energy
Ni-Cd D size	25 Ah/kg 72 Ah/l	1.2V	30 Wh/kg 87 Wh/l
Ni−MH[1] AA size	40 Ah/kg 119 Ah/l	1.26V	51 Wh/kg 150 Wh/l
Sub-C size	50 Ah/kg 168 Ah/l	1.3V	65 Wh/kg 219 Wh/l
LiCoO$_2$ − C[2] AA size	22 Ah/kg 53 Ah/l	3.6V	78 Wh/kg 192 Wh/l
D size	32 Ah/kg 70 Ah/l	3.6V	115 Wh/kg 253 Wh/l
LiNiO$_2$ − C[3] (33%)	25 Ah/kg	3.2V	80 Wh/kg
LiMn$_2$O$_4$-C (33%)	23 Ah/kg 44 Ah/l	3.7V	83 Wh/kg 163 Wh/l
Li$_{1.1}$Mn$_2$O$_4$-C (33%)	25 Ah/kg	3.7V	91 Wh/kg
Li$_2$Mn$_2$O$_4$-C (33%)	35 Ah/kg	2.9V	100 Wh/kg

[1] from (18). [2] from (10). [3] calculated from data taken with coin cells (19).

124

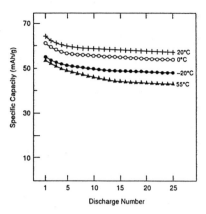

Figure 6: Capacity versus cycle number for $LiMn_2O_4/EC + DEE(50:50) + 1MLiClO_4/C$ (petroleum coke) cells as a function of the operating temperature. The cells have been cycled at a rate of C/2 in the 4.35V-2.3V voltage range at room temperature, C/2 in the 4.1V-2.3V range at 55°C and C/5 in the 4.35V-2V range at 0°C and -20°C.

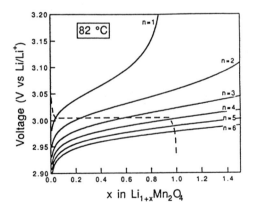

Figure 7: A calculation of Nernst equation at 82°C for the reaction: $LiMn_2O_4 + 3x/2\ LiI ---> Li_{1+x}Mn_2O_4 + x/2\ LiI_3$ has been done in order to determine the amount of LiI needed to prepare $Li_{1+x}Mn_2O_4$ powders of any wanted lithium excess (x). The amount of $LiMn_2O_4$ has been fixed at 4g, while nX4g of LiI is used in $100cm^3$ of acetonitrile. The dashed curve represents the voltage of the $LiMn_2O_4/Li_2Mn_2O_4$ phase transformation measured at 82°C.

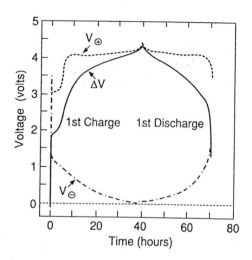

Figure 8: The first cycles at room temperature of a $Li_{1.1}Mn_2O_4/$ $EC + DEE(50:50) + 1MLiClO_4/C$ (petroleum coke) cell cycled between 4.35 V and 2.3 V, as measured with the three-electrode experiment.

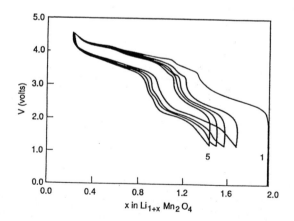

Figure 9: Cycling behavior at room temperature of a $Li_2Mn_2O_4/$ $EC + DEE(50:50) + 1MLiClO_4/C$ rocking-chair cell at a rate of C/10.

RECHARGEABLE LITHIUM BATTERIES

USING V_6O_{13}/V_2O_5 AS THE POSITIVE ELECTRODE MATERIAL

René Koksbang

Innocell Aps
Hestehaven 21 D, DK-5260 Odense S, Denmark

Pascal Gé

Innovision A/S
Munkebjergvænget 13, DK-5230 Odense S, Denmark

ABSTRACT

Positive electrodes containing a mixture of V_6O_{13} and α-V_2O_5 is evaluated and compared to other positive electrode materials. Cycling of both materials is wellknown and a mixture provide a higher average voltage and a less sloping voltage curve. The energy density, even at shallow cycling, is competitive to other well characterized active cathode materials exploited for use in secondary lithium batteries.

INTRODUCTION

Although insertion compounds have been known as lithium ion host materials for rechargeable lithium batteries for more than a decade [1], the search for new compounds has not revealed any with a flat discharge curve close to 3 V vs Li. All the known compounds have sloping discharge curves and most of the curves have plateaus characteristic of the host material. Recent progress has focussed on rechargeable manganese oxide compounds [2-5]. For these compounds the lithium insertion reaction proceed as a displacement reaction close to 2.8 V vs Li. However, only a fraction of the capacity achieved with the primary MnO_2 systems can be utilized in the secondary systems.

In this paper we describe another route to the achievement of a rechargeable lithium battery with a high voltage, close to 3 V vs lithium, and a competitive energy density, by the use of a mixture of active electrode materials. By proper selection of the materials, it is possible to realize average voltages close to 3 V vs Li.

A similar approach has previously been used to make high energy and high rate

127

solid cathodes [6] by Whittingham et. al. They used a combination of e.g. TiS_2 and α-MoS_3, where TiS_2 provided the high rate (high electronic conductivity) and α-MoS_3 provided the high energy. The theoretical energy density of α-MoS_3 is 2-3 times that of TiS_2. This combination of materials provided a cathode with a voltage close to 2.2 V. Recently, the approach has also been used to improve the unsatisfactory energy content of rechargeable Li_xMnO_2 cathodes [7], by manufacture of manganese/vanadium oxide compounds/mixtures. As will be shown, the energy content of this material is inferior to that of the V_2O_5/V_6O_{13} combination.

EXPERIMENTAL

V_2O_5 (Gesellschaft für Electrometallurgie) was used as-received. V_6O_{13} was prepared in-house by thermal decomposition of NH_4VO_3 (Gesellschaft für Electrometallurgie) under a nitrogen flow [9].

The cathodes contained Shawinigan black, active material and a binder in the weight ratio 10:60:30 and was prepared by pressing cathode sheets under a pressure of 3 tons/cm^2 as described elsewhere [10,11].

Propylene carbonate (PC, Hüls GmbH), was distilled over sodium at reduced pressure. The first 100 ml fraction was discarded and the rest transferred to an Ar filled glovebox. $LiCF_3SO_3$ (FC-124, 3M) was dried by heating to 120°C in vacuum overnight. Li foil (Cypruss Foote Mineral), surface cleaned by scraping and rolling was used as the anodes.

The cells were assembled in either a glovebox filled with Ar (moisture level below 10 ppm) or in a dry room with a relative humidity below 2 %. The cells were placed in stainless steel button cells sealed by a polypropylene seal. The water content of electrolytes and other materials was controlled by Karl Fischer titration and was in general lower than 50 ppm.

The cells were cycled galvanostatically between preset voltage limits with current densities of 0.1 mA/cm^2 in both discharge and charge. After galvanostatic charge, the cells were kept potentiostatically at 3.5 V for one hour. The utilization of the active material was estimated from the cathode compositions, the masses of the cathodes and the amount of charge passed through the cells.

RESULTS AND DISCUSSION

During lithium insertion in V_2O_5, the largest fraction of the capacity is obtained at plateaus close to 3.2 V vs Li (1 Li/V_2O_5) and at ca. 2.1 V vs Li (Fig. 1). On the other hand, lithium insertion in V_6O_{13} takes place at three voltage plateaus close to 2.7, 2.5 and 2.2 V vs Li respectively (Fig. 1). These two materials are therefore suitable for use in a combination cathode containing both materials, because the voltage window of the Li/V_6O_{13} couple "fit" into the large voltage drop which separate the two main plateaus of the Li/V_2O_5 couple. Displacement of the V_2O_5/volt curve along the horizontal axis, show the general shape expected from a mixed system as depicted in Fig. 2. The figure is constructed by the displacement of the Li/V_6O_{13} curve in Fig. 1 as described above. Figs. 1 and 2 also show that the optimum ratio between the two compounds is 50 molar-% of each compound, because this corresponds to 50 % utilization of the active material at 50 % DOD.

Comparison between actual voltage curves of the Li/V_6O_{13}, Li/V_2O_5 and $Li/V_6O_{13}/V_2O_5$ (Fig. 2 and 3) systems indeed show that the shape of the voltage curve of the latter system correspond to a superimposition of the two former curves and that the average voltage is higher for the combined system, although the characteristic plateaus of the curve shown in Fig. 2. is less pronounced than in the constructed curve.

It has previously been shown that both compounds can be cycled reversibly in lithium cells [8-9], especially at the low utilization of the active materials. Because the volume changes associated with the lithium insertion processes are smaller than by full utilization of either of the materials, the stress build-up in the electrodes is correspondingly smaller, and consequently the maintenance of the mechanical integrity of the electrode is improved also.

Examples of the cycling behavior of such cells to various depths of discharge, are shown in Fig. 4. Because the experiments only aimed at demonstrating the principle, the cycling was terminated after 25 cycles. During cycling to the lower voltage limits, 1.5 and 1.8 V, high capacities are achieved but at the expense of a decaying capacity. During shallow cycling, the capacity retention is excellent, without any significant capacity loss after the first few cycles, for shallow cycling of the cathodes. The energy density is similar to, or greater than that of TiS_2 based systems. Furthermore, whereas overdischarge will severely impart the cyclability of e.g. TiS_2 because of irreversible structural changes of the host material, the V_2O_5/V_6O_{13} cathode has an excess capacity which can provided higher energy without significant loss of the capacity (Fig. 4) on continued cycling. The additional capacity thus act as an overdischarge protection.

In Tab. 1, the theoretical specific energy of the V_2O_5/V_6O_{13} at different depths of discharge is compared to other active cathode materials. At very shallow cycling (3.0-3.5 V) this system is inferior to any other material or combination of materials. However, by cycling between 2.5 and 3.5 V, i.e. corresponding to 50 % DOD of each of the two materials, the specific energy is competitive to TiS_2 at a higher average voltage. Considering that the V_2O_5/MnO_2 system [7] cited in the table, is discharged to 1.0 V, the V_2O_5/V_6O_{13} compare favorably with this system also when the former is discharged to 2.5 V vs Li also as this only provide ca. 150 Ah/kg and 450 Wh/kg [7].

CONCLUSION

Cathodes containing properly selected insertion compounds such as V_2O_5 and V_6O_{13} has an energy content which is competitive to some of the most studied transition metal oxides, including the rechargeable manganese oxides, and sulphides which are under consideration for use in rechargeable lithium batteries.

Furthermore, this is achieved at an average voltage close to 3 V vs lithium and during shallow cycling of the cells. The cycling performance appear to be similar to the Li/TiS$_2$ system as no appreciable capacity loss was experienced during cycling under the conditions applied in this work.

Manganese dioxides are lowered priced than most other transition metal compounds which are possible candidates for use as cathode materials in rechargeable lithium batteries. However, the cost difference is somewhat alleviated by the necessary post treatment with e.g. LiNO$_3$, LiOH or n-BuLi and heating, which is needed for optimum cycling and energy content.

REFERENCES

1. J. DeSilvestro, O. Haas, *J. Electrochem. Soc.*, 137, 5C (1990)
2. L. Li, G. Pistoia, *Solid State Ionics*, 47, 231 (1991)
3. '.. Li, G. Pistoia, *Solid State Ionics*, 47, 241 (1991)
4. J. M. Tarascon, E. Wang, F. K. Shokoohi, W. R. McKinnon, S. Colson, *J. Electrochem. Soc.*, 138, 2859 (1991)
5. J. M. Tarascon, D. Guyomard, *J. Electrochem. Soc.*, 138, 2864 (1991)
6. M. S. Whittingham, A. J. Jacobson, *J. Electrochem. Soc.*, 128, 485 (1981)
7. N. Kumagai, S. Tanifuji, K. Tanno, *J. Power Sources*, 35, 313 (1991)
8. K. West, B. Zachau-Christiansen, M. J. L Østergård, T. Jacobsen, *J. Power Sources*, 20, 165 (1987)
9. K. West, B. Zachau-Christiansen, T. Jacobsen, S. Atlung, *J. Power Sources*, 14, 235 (1985)
10. R. Koksbang, D. Fauteux, P. Norby, K. A. Nielsen, *J. Electrochem. Soc.*, 136, 598 (1989)
11. R. Koksbang, P. Norby, *Electrochim. Acta*, 36, 127 (1991)

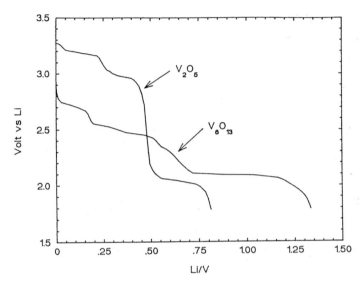

Fig. 1. Voltage for lithium insertion in V_2O_5 and V_6O_{13}.

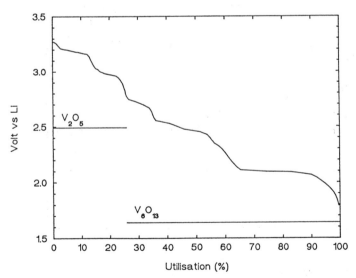

Fig. 2. Expected shape of voltage curve for a cathode containing mixed V_2O_5/V_6O_{13} active material.

131

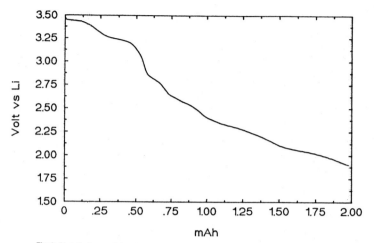

Fig. 3. First discharge of the mixed V_2O_5/V_6O_{13} cathode. The V_2O_5/V_6O_{13} ratio is 50:50 mol-% and the current density 0.1 mA/cm².

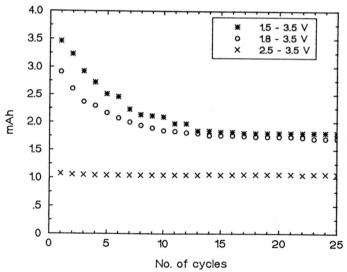

Fig. 4. Cycling curves as function of depth of discharge. The lower voltage levels are indicated in the figure. The current densities were 0.1 mA/cm² in both charge and discharge.

Tab. 1. Comparison of the specific energies for different active cathode materials.

Material	x	y	Volt	Ah/kg	Wh/kg
Li_xTiS_2	1.0	0.0	1.7-2.4	239	490
$Li_xV_6O_{13}$	8.0	0.0	1.8-3.0	377	905
$Li_xV_2O_5$	2.0	0.0	2.0-3.5	295	811
Li_xMnO_2	0.5	0.0	2.0-3.6	154	431
*$Li_xV_2O_5$/ Li_yMnO_2	---	---	1.0-3.5	340	765
$Li_xV_2O_5$/ $Li_yV_6O_{13}$	1.0	0.0	3.0-3.5	74	241
$Li_xV_2O_5$/ $Li_yV_6O_{13}$	1.0	4.0	2.5-3.5	178	534
$Li_xV_2O_5$/ $Li_yV_6O_{13}$	2.0	8.0	1.8-3.5	356	943

*V/Mn ratio = 1, discharge to 1 V vs Li [7].

ELECTROCHEMICAL STABILITY OF ORGANIC ELECTROLYTES IN CONTACT WITH SOLID INORGANIC CATHODE MATERIALS

P. Ge
Innovision A/S, Hestehaven 21 D
5260 Odense S, Denmark

R. Koksbang*
Innocell Aps, Hestehaven 21 D
5260 Odense S, Denmark

ABSTRACT

The oxidation stability of propylene carbonate based, organic electrolytes, in contact with solid cathode materials was investigated, and compared to the cycling performance of the corresponding batteries. The cathode materials were chosen to represent different chemistries (oxides and sulphides) and a wide potential range (1.0 - 3.5 V) for reversible lithium insertion.

INTRODUCTION

In the course of investigation of insertion compounds for use as positive electrode materials in rechargeable lithium batteries, the majority of the research have been devoted to the amount of lithium the compounds are able to accommodate and to the structural stability of the compounds. Thus, the major effort has been placed on the lower potential limit of lithium batteries incorporating insertion compound cathodes, whereas the upper potential limit usually is slightly higher than the OCV of a new cell. Very little attention has been directed towards the stability of the organic electrolyte in contact with the actual cathodes at high potentials. Usually the electrolyte stability is investigated by the use of noble metal (Pt, Au) or glassy carbon (GC) electrodes. However, lithium battery cathodes are usually constructed as porous electrodes containing an active material (transition metal oxides and sulphides), a binder and in most cases graphite or acetylene black type carbon. Not only have the electrodes a surface area, which exceed the geometrical area significantly because small particle size powders are used, but the materials used in the electrodes have a higher chemical

---

*Author to whom correspondance should be adressed.

activity than the commonly used noble metals as is well known from the catalytic activity of these materials. V_2O_5 is e.g. used as oxidation catalyst for the oxidation of hydrocarbons [1].

Evidence for formation of an interface layer on the cathode have been provided by A.C. impedance analysis [2-4]. In this case, the most studied compound was V_6O_{13} but other materials have been under study as well with similar results. The formation of an interface layer suggest that a reaction between the cathode components and the electrolyte components take place.

As a fast mean to estimate the electrochemical stability of electrolytes, voltammetric methods have been widely used. For this purpose, inert working electrodes, such as Au, Pt, glassy carbon (GC) is often used [5 and references therein]. Ossola et. al used linear sweep voltammetry for comparison of the anodic electrolyte stability of different electrolytes in contact with Pt, teflonized carbon (TAB), GC and LiV_3O_8 [5]. The measurements showed that the anodic potential stability limit was independent of the working electrode material. On the other hand, OCV vs time recordings of LiV_3O_8 based cells [5] showed that the OCV was declining, at least at potentials above 3 V indicating that electrolyte oxidation takes place.

Due to the scarce data available in the literature, we decided to determine the anodic stability potential limit (E_{ox}), of PC based electrolytes in contact with different cathode materials. The cathode materials chosen, represent a wide potential window for reversible lithium insertion (Tab. 1) and different chemistries (sulphides vs oxides). These results are compared to actual cycling data based on identical materials.

EXPERIMENTAL

Solvents: Propylene carbonate (PC) (Hüls GmbH), was distilled over sodium at reduced pressure. The first 100 ml fraction was discarded and the rest transferred to an Ar filled glovebox. The high purity cosolvents (Merck, p.a.), dimethylformamide (DMF), gamma-butyrolactone (BL) and ethylene carbonate (EC) were used as received except for drying over molecular sieves.

Salts: $LiCF_3SO_3$ (3M), $LiPF_6$ and $LiClO_4$ (Aldrich) were dried by heating to 120°C in vacuum overnight. $LiAsF_6$ (Lithco) was used as-received. 1 M salt solutions were used throughout.

Anode: Li foil (Cypruss Foote Mineral), surface cleaned by scraping and rolling was used as the anodes.

Cathode materials: MoO_2 (Alfa Ventron) and V_2O_5 (Gesellschaft für Electrometallurgie), were used as-received. V_6O_{13} was prepared in-house by thermal decomposition of NH_4VO_3 (Gesellschaft für Electrometallurgie) under a nitrogen flow [6]. α-MoS_3 was prepared by acidic precipitation from an aqueous solution of $(NH_4)_2MoS_4$ (Alfa Ventron) [7]. The phase purity of the crystalline materials was controlled by powder X-ray diffraction and comparison with the appropriate JCPDS card

135

files (MoO_2: 32-671, V_6O_{13}: 27-1318, V_2O_5: 9-387). MoO_2 and V_2O_5 were both pure phases whereas the V_6O_{13} sample contained a small amount of V_2O_5. The α-MoS_3 sample was X-ray amorphous.

The water content of electrolytes and other materials was controlled by Karl Fischer titration and was in general lower than 50 ppm.

Cathodes: The cathodes contained Shawinigan black, active material and a PTFE binder in the volume ratios 45/45/10. The electrodes were prepared by pressing cathode sheets under a pressure of 3 tons/cm^2 as described elsewhere [8].

Reference electrodes: Reference electrodes were incorporated in the cells as described elsewhere [8] and consisted of either a strip of lithium foil or LiAl foil.

Cell assembly: The cells (Fig. 1) were assembled in either a glovebox filled with Ar (moisture level below 10 ppm.) or in a dry room with a relative humidity below 2 %. The cells used for the cycling experiments, were placed in stainless steel button cells sealed with a polypropylene seal. Only 4-5 drops of electrolyte was added to the cells which were therefore essentially starved. The cells used for the cyclic voltammetry measurements were flooded and is described elsewhere [8].

Cyclic voltammetry: The sweep rate was in general 50 mV s^{-1} and multiple sweep were performed between different voltage limits. The measurements were carried out using a Solartron 1286 Electrochemical Interface and a Thompson potentiostat equipped with a ramp generator. Electrolyte oxidation was considered to occur when the current density exceeded 0.1 mA/cm^2. This value corresponds to approximately 10 % of a typical current density applied to this type of lithium batteries. In the following, the potential at which the current density reach 0.1 mA/cm^2 is denoted E_{ox}.

Cell cycling: The cells were cycled galvanostatically between preset voltage limits with current densities of 0.1 mA/cm^2. After discharge the cells were left at OCV for one hour before recharge was initiated. Similarly, after the galvanostatic charge, the cells were kept potentiostatically at the upper voltage limit for one hour prior to discharge. The utilization of the active material was estimated from the cathode compositions, the masses of the cathodes and the amount of charge passed through the cells.

RESULTS AND DISCUSSION

The cathode materials which have been used for these studies, MoO_2, V_6O_{13}, V_2O_5 and α-MoS_3, are able to accommodate respectively 1, 8, 2 and 3 lithium ions per formula unit reversibly, in the potential ranges indicated in Tab. 1. When the potential limits shown in Tab. 1 are exceeded, a possible electrochemical reaction will involve the electrolyte, i.e. solvent, salt or both. These reactions are irreversible and will have a detrimental effect on the battery performance due to the degradation of the electrolyte components.

In the following, the experimental results will be primarily exemplified using MoO_2 as the cathode material and $LiCF_3SO_3$ as the electrolyte salt. The other salts have not been fully examined but has merely been used for control experiments.

In Fig. 2, the first cathodic sweep at low rate, of a cyclic voltammogram of a MoO_2 containing cathode, is shown. The peaks A and B correspond to the reversible lithium insertion reaction which is observed as two plateaus during galvanostatic cycling of MoO_2 based lithium batteries (Fig. 3). Apart from this reaction, a strong oxidation peak, corresponding to electrolyte oxidation, is observed above ca. 3.7 V vs Li. Though a cathodic sweep was applied in this case, the sweep was generally started anodically in order to avoid confusion between the currents caused by the deinsertion and the electrolyte oxidation current.

As illustrated in Fig. 4A, the current arising from the deinsertion reaction contribute significantly to the total current and therefore make it difficult to estimate the electrolyte oxidation current. From this figure, the E_{ox} for 1 M $LiCF_3SO_3$ in pure PC was estimated to be 3.7 V in the first sweep whereas it was impossible to estimate the E_{ox} after the second sweep. Note that E_{ox} is defined as the potential at which the current density exceed 0.1 mA/cm^2.

In Figs. 4B-4D, voltammograms similar to that of Fig. 4A are shown. The three samples used for these experiments, each contained 50 v/o cosolvent (EC, BL and DMF). Evidently, the voltammograms are almost identical to the one shown in Fig. 4A which show that no improvement is achieved by addition of these solvents.

Linear sweep voltammograms of V_6O_{13} based cathodes in three different PC based electrolytes containing 1 M $LiCF_3SO_3$, are shown in Fig. 5. Apart from the DMF containing electrolyte, for which oxidation starts at 4.0 V, the stability limit exceed 4.1-4.2 V.

By comparison with MoO_2 in identical electrolytes, this is somewhat surprising, considering the strongly oxidizing properties of vanadium(V) containing oxides [1]. This may be explained by the difference in overpotential, i.e. the potential difference between end of the deinsertion reaction and initiation of electrolyte oxidation, rather than the absolute potential. The overpotential of MoO_2 based cells is ca. 1.7 V (2.0-3.7 V, Tab. 1 and 2), whereas the value is only 0.7 V and 1-1.2 V for V_2O_5 and V_6O_{13}, respectively.

Tab. 2 summarise the results obtained with MoO_2 and similar measurements with other active cathode materials. The stability of PC based electrolytes appear to be independent of the choice of solvent composition. On the other hand, value of E_{ox} seem to depend on the cathode material and is decreasing in the order:

$$V_2O_5 > V_6O_{13} > MoO_2$$

The influence of the Shawinigan black is considered to be negligible compared to the transition metal oxides because the carbon content, on a volumetric basis, is almost identical in all the electrodes. Likewise, no significant variations in E_{ox} were observed by substitution of $LiCF_3SO_3$ with $LiAsF_6$, $LiPF_6$ and $LiClO_4$. This is presumably related to an E_{ox} for the salts which is higher than the E_{ox} of the solvents. The experiments with α-MoS_3 showed no significant differences between oxides and sulphides up to 3.5 V.

137

The E_{ox} is lower than measured on Pt and GC electrodes. Although this could be attributed to the larger surface area of the composite cathode, this is unlikely because of the variations in E_{ox} observed for different cathode materials. It is therefore not possible to estimate the oxidation stability of organic electrolytes from cyclic voltammetry on inert electrodes.

Cycling results are shown for V_2O_5 based button cells in Fig. 6. These cells were all cycled between 2.8 and 3.8 V vs Li, i.e. with a lower voltage limit well within the reversible lithium insertion range (Tab. 1), and to an upper voltage limit which is significantly lower than the E_{ox} but higher than the voltage corresponding to complete removal of Li from $Li_xV_2O_5$. It may be noted that although E_{ox} is identical for these systems, the cyclability is significantly different. As the Li content in the anodes exceeded the cathode capacity by approximately a factor of 30, the capacity loss is considered to occur because of oxidative electrolyte break down on the cathode.

The end of cycle life is reached suddenly and examination of the cells after cycling revealed that the electrolyte contained brown reaction products. No attempts have been made to identify these reaction products.

Similar results were obtained with α-MoS_3 and V_6O_{13} containing cathodes in contact with these electrolytes. There is thus no correlation between E_{ox}, determined by cyclic voltammetry, and cycling of actual batteries.

The cycling results are consistent with cycling results for MoS_2 cells [12] which were cycled to upper voltage limits in the range 2.0 - 2.4 V. The Li/MoS_2 couple is fully charged at 2.0 V. At higher voltages the number of attainable cycles decreased drastically.

CONCLUSION

The major conclusions which can be drawn from the cyclic voltammograms are 1) the potential (E_{ox}) at which electrolyte oxidation occur, depends on the cathode material used in the composite cathode and is decreasing in the order $V_2O_5 > V_6O_{13} > MoO_2$, 2) the oxidation stability of PC based electrolytes, is not improved by addition of EC, BL or DMF as cosolvents, 3) the E_{ox} is lower than measured on Pt and GC electrodes. It is therefore not possible to estimate the oxidation stability of organic electrolytes from cyclic voltammetry on inert electrodes.

The electrolytes appear to be stable to at least 3.5 V when in contact with α-MoS_3 containing composite cathodes.

E_{ox} appear to be independent of the nature of the electrolyte salt and of the carbon in the electrodes.

Finally, cells recharged to potentials in between E_{ox} and the OCV of fresh cells, show an appreciable capacity loss during cycling and the cycle life is usually terminated abruptly, in spite of the lack of dendrites. The cycling performance depends on the solvent combination used but appear not to be related to the oxidation stability potential as determined by cyclic voltammetry.

Thus, voltammetric methods can only be used for preliminary measurements at best. The estimation of the oxidation stability of the electrolytes by these techniques therefore seem to be of little practical value.

ACKNOWLEDGEMENT

Dr. K. West, The Technical University of Denmark, is thanked for provision of the sample of α-MoS$_3$ used for the experiments.

REFERENCES

1. T. Ono, Y. Nakagawa, Y. Kubokawa, *Bull. Chem. Soc. Jpn.*, 54, 343 (1981)
2. P. G. Bruce, F. Krok, *Electrochim. Acta*, 33, 1669 (1988)
3. P. G. Bruce, F. Krok, *Solid State Ionics*, 36, 171 (1989)
4. P. G. bruce, E. McGregor, C. A. Vincent in B. Scrosati (ed.) "*Second International Symposium on Polymer Electrolytes*", Elsevier Applied Science, London, 1990, pp. 357.
5. F. Ossola, G. Pistoia, R. Seeber, P. Ugo, *Electrochim. Acta*, 33, 47 (1988)
6. K. West, B. Zachau-Christiansen, T. Jacobsen, *Electrochim. Acta*, 28, 1829 (1983)
7. J. J. Auborn, Y. L. Barberio, K. J. Hanson, D. M. Schleich, M. J. Martin, *J. Electrochem. Soc.*, 134, 580 (1987)
8. R. Koksbang, D. Fauteux, P. Norby, K. A. Nielsen, *J. Electrochem. Soc.*, 136, 598 (1989)
9. D. W. Murphy, F. J. DiSalvo, J. N. Carides, J. V. Waszczak, *Mat. Res. Bull.* 13, 1395 (1978)
10. K. West, B. Zachau-Christiansen, M. J. L. Østergård, T. Jacobsen, *J. Power Sources*, 20, 165 (1987)
11. S. R. Narayanan, S. Surampudi, A. I. Attia, in (K. M. Abraham, M. Salomon, eds.), *Proceeding of the Symposium on Primary and Secondary Lithium Batteries*, Vol. 91-3, The Electrochemical Society, Inc., Pennington, NJ, USA, 1991, pp. 109.
12. K. Kumai, K. Ishihara, Th. Ikeya, T. Iwahori, T. Tanaka, *ibid.*, pp. 371.

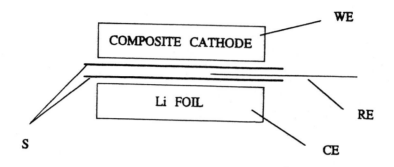

Fig. 1. Cell assembly: WE = composite cathode, CE = Li foil anode, RE = Li or LiAl foil reference electrode and S = Celgard 2400 polypropylene separator.

Tab. 1. Potential range for reversible insertion of Li

Material	Li/formula unit	Potential range	Reference
MoO_2	1	1.0 - 2.0 V	9
V_6O_{13}	8	1.8 - 3.0 V	6
V_2O_5	2	2.0 - 3.5 V	10
α-MoS_3	3 (4)	1.0 - 2.5 V	7

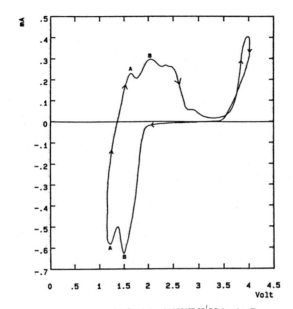

Fig. 2. Cyclic voltammogram of MoO_2 cathode and 1 M $LiCF_3SO_3$/PC electrolyte. First sweep at 0.1 mV/s cathodically.

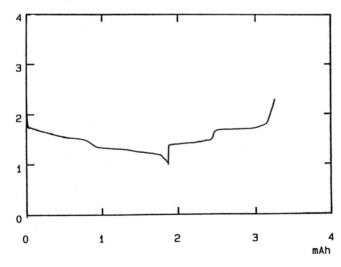

Fig. 3. Galvanostatic discharge of a MoO_2 cathode.

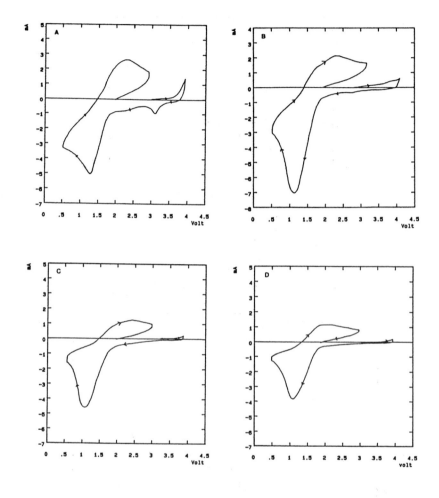

Fig. 4. Cyclic voltammogram of MoO_2 cathode and 1 M $LiCF_3SO_3$ electrolyte. First sweep at 50 mV/s anodically. A) PC/DMF, B) PC/BL, C) PC/EC and D) PC. The electrolytes contained solvent mixtures in the volume ratio 1:1.

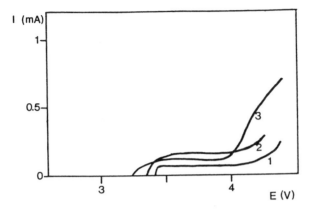

Fig. 5. Linear sweep voltammograms of V_6O_{13}, at 50 mV/s anodically. The electrolytes are 1 M LiCF$_3$SO$_3$ in 1) PC/EC, 2) PC and 3) PC/DMF.

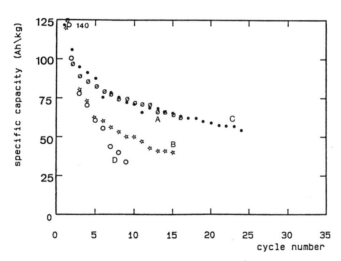

Fig. 6. Cycling of V_2O_5 cells between 2.8 and 3.8 V at 0.1 mA/cm^2. Electrolytes: A) PC, B) PC/EC, C) PC/DME and D) PC/BL. Salt: 1 M LiCF$_3$SO$_3$.

Tab. 2. E_{ox} (volt) for various electrolytes in contact with different cathode materials. The electrolyte salt was $LiCF_3SO_3$ (1 M), and the solvent mixtures were 1:1 in volume ratio.

Cathode material	E_{ox} PC	E_{ox} PC/EC	E_{ox} PC/DMF	E_{ox} PC/BL
MoO_2	3.7	3.7	3.7	3.7
V_6O_{13}	4.15	4.15	4.0	4.1
V_2O_5	>4.2	>4.2	>4.2	---
α-MoS_3	>3.5	---	---	---
Pt	4.7*	---	---	---
$Li_xV_3O_8$	4.6*	---	---	---
TAB	4.7*	---	---	---

*PC/DME (1:1), $LiAsF_6$ and $LiClO_4$ [5].

Stabilization and Improvement of Li Battery Electrolyte Solutions Based on
1-3 Dioxolane

Y. Gofer, M. Ben-Zion and D. Aurbach
Department of Chemistry
Bar-Ilan University
Ramat Gan 52900, Israel

Abstract

Highly stable solutions of 1-3-dioxolane (DN) with $LiClO_4$ or $LiAsF_6$ may be prepared by the use of tertiary amine additives. Very high Li cycling efficiency is obtained with stabilized $DN/LiAsF_6$ solutions. These electrolytes can be further improved by addition of alkyl carbonates as co-solvents. The correlation between Li cycling efficiency and Li surface chemistry in these systems was investigated using surface sensitive FTIR and SEM - X-ray microanalysis techniques.

Introduction

1-3 Dioxolane (DN) was listed in the past among the most promising solvents for rechargeable lithium batteries.[1] Quite a high conductivity is obtained with Li salt solutions in DN at high concentration (1-2M), and very good cycling efficiency was obtained in half cell testing of Li electrodes in $DN/LiClO_4$ solutions.[2] Other $LiClO_4$ solutions in most polar aprotic solvents, including ether, alkyl carbonate, lactones and esters, gave very poor performance in Li half cell testing experiments.

However, the interest in DN as a primary solvent for secondary lithium batteries declined considerably after a few explosions occurred during laboratory testing of Li cells containing $LiClO_4/DN$ solutions.[3] It seems that this solvent's main problems are its tendency to polymerize. A sudden polymerization could explain the change in solutions that led to the explosion of Li cells containing DN. While $LiClO_4/DN$ solutions are usually stable, $LiAsF_6/DN$ solutions polymerize readily, probably due to an unavoidable presence of Lewis acids such as trace AsF_3. This excludes the possibility of using $LiAsF_6$, which is a superior electrolyte for secondary Li systems, with DN without a special stabilization.

In this work, DN solutions of $LiClO_4$ and $LiAsF_6$ stabilized by tertiary amines were investigated as electrolyte solutions for secondary Li batteries. Tertiary amines are potential stabilizing agents since as bases they can neutralize Lewis acids in solutions and hence avoid DN polymerization. A variety of T-amines were tested. The list includes aliphatic amines such as triethyl, tripropyl and tributyl amines, aromatic amines such as tribenzyl amine and cyclic amines such as N-methyl piperidine. Possible reaction of the amines with lithium, both neat and in DN solution, were investigated using surface sensitive FTIR and X-ray microanalysis. Li cycling efficiency in half cell testing using T-amine stabilized DN solutions was measured under various conditions. The electrochemical window and the degree of stabilization of DN solutions containing T-amines was explored using cyclic voltammetry (with noble metal electrodes), as partly polymerized or pure DN solutions have characteristic voltammetric behavior.[4] Scanning electron microscopy (SEM) was used in order to correlate between the morphology and the cycling efficiency of the Li electrodes under the various experimental conditions.

In addition, several reports describing EC-ether mixtures containing $LiAsF_6$ as very good electrolyte solutions for secondary Li batteries[5] led us to investigate $DN/LiAsF_6$ solutions containing ethylene carbonate (EC) or propylene carbonate (PC) as cosolvents.

Experimental

DN (Aldrich) was triply distilled over sodium and benzophenone under argon (1 atm). T-amines (Aldrich) were also distilled under argon before use. $LiClO_4$ (Aldrich) was dehydrated in a vacuum oven (10^{-2} - 10^{-3} mm/Hg, 150°C 3 days). $LiAsF_6$ (Lithco) was used as received. PC

145

(Aldrich) and EC (Aldrich) were vacuum distilled over CaH_2 (in order to obtain the proper pressure, which was about 6-8 mm/Hg, high purity argon was used). All the electrochemical measurements and the preparation for the spectroscopic measurements were performed under high purity argon in glove box systems and were described previously .[6-9] The FTIR, and the SEM measurements of lithium or non-active metal (mostly nickel) surfaces and the transfer techniques from the glove boxes to the spectrometers were described elsewhere.[6,7] Conductivity measurements were performed using an El Hama TH-2400 conductometer. Lithium cycling efficiency was measured both in parallel plate cells where the electrodes are separated by a few mm of solution, described elsewhere[7,10] and in the cell shown in figure 1 where the electrodes are closed and pressed together, separated by a 2400 Celgrad polypropylene separator. In both types of cells the working electrodes were lithium deposited on nickel or copper substrates from the electrolyte solution in the cell (5-20 C/cm^2). The counter electrodes were lithium foils. In a typical experiment, a portion of the deposited lithium (DOD ranged from 10-80%) was charged-discharged consecutively several times (number of cycles was 5-100, depending on the experiment), galvanostatically. (Current density ranged from 0.5 - 10 mA/cm^2). Li cycling efficiency was calculated from the residual active lithium left after 'the cycling, as previously described.[7]

Results and Discussion

a. T-amines as stabilizers for DN solutions and their compatibility with lithium

As was previously reported, $LiClO_4$/DN solutions may partly polymerize due to storage over molecular sieve, activated alumina or contamination with trace Lewis acids, and at potentials above 3 - 3.2V (Li/Li^+).[4] The voltammetric behavior of partly polymerized solutions with noble metal electrodes is characterized by high cathodic currents below 1.5V (Li/Li^+) because of a breakdown of the natural electrode passivation at low potential due to interaction of the electrode passive films with the polymerization products, as already discussed.[4]

Figure 2a (solid line) presents a typical cyclic voltammogram of partly polymerized $LiClO_4$/DN solvent (Au electrode). Polymerization was initiated in this case by applying potentials above 3.2V to this system. This voltammogram is indeed characterized by high cathodic currents at potentials below 1.5V, even in the anodic sweep. Addition of trace tertiary amines to these solutions seems to prevent this polymerization. Figure 2a (dashed line) shows a voltammogram obtained with DN/$LiClO_4$ solution containing tribenzyl amine (TBzA). Although the voltage upper limit in this case was 4V (Li/Li^+), which is $\approx$ 1V above DN polymerization potentials,[6,8] this voltammogram (Fig 2a, dashed line) is typical of unpolymerized solution. The increase in current around 4V is due to the amine oxidation.

A similar stabilization effect was observed when other T-amines were added, even in trace amounts (>100 ppm), to DN/$LiClO_4$ solutions. The list include trialkyl amines R = ethyl, propyl butyl and octyl, triaryl amines, R = benzyl, phenyl, 1-methyl piperidine and 1-methyl morpholine.

While $LiAsF_6$/DN solutions readily polymerize, those containing the above T-amines are stable for years. Figure 2b shows a typical voltammogram obtained from DN/$LiAsF_6$ 1M solutions stabilized with TEA (1000 ppm). This voltammogram is typical of those obtained with Li salt solutions in all other polar aprotic solvents (e.g. ethers, alkyl carbonates,[8,9]) and therefore, proves that the DN/$LiAsF_6$ solution is stable and unpolymerized in spite of the voltages applied which readily induce polymerization of unstabilized DN solutions. As shown in Figure 3c, the electrochemical window of Li salt solutions stabilized by T-amines is limited to 3.2 - 3.7 V, depending on the amine used. Although this window is narrower than that of other Li battery electrolyte solutions, it is still compatible with cathodes such as TiS_2.

The compatibility of T-amines as DN stabilizers with lithium electrodes was rigorously studied using surface sensitive techniques. Figure 3 compares FTIR spectra obtained from the

146

reaction products of Li amalgam with propylamine and tripropylamine and from Li surface stored for a few days in tripropylamine. The T-amine was previously cleaned either by storage over Na-K or Hg/Li amalgams in order to eliminate any protic contaminants. As shown in this figure, all the spectra are very similar and typical of a mixture of propylamine and dipropylamine.[11] The spectrum obtained from the Li surface (Figure 3, lower spectrum), also have a Li nitride (Li_3N) peak around 680 cm^{-1}.

Hence, these studies prove that lithium or Li/amalgams react with T-amines to form three types of amides, R_2NLi, $RNLi_2$ and NLi_3. Other products are R_2 or RH. (Gas evolution was observed during storage of T-amines over Li/Hg amalgam). The fact that the spectra of Figure 3 are typical of amines should be attributed to an unavoidable reaction of the amides formed with trace water and the LiOH peak around 3680 cm^{-1} shown in part of these spectra further proves it.

However, in spite of the reactivity of T-amines with lithium, it would seem that as trace amounts in DN solutions they do not react with lithium. Figure 4 compares FTIR spectra obtained *ex situ* from Li samples stored for a few weeks in DN-T amine solutions. A spectrum obtained from Li surface stored in pure DN is also shown (Figure 4a). All these spectra are similar and have peaks around 2950-2800, 1100-1000, 850, 800 and 600-500 cm^{-1}, which should be attributed to Li alkoxides (ROLi).[10] As indicated in Figure 4, all of these spectra have peaks around 1500-1450 and 870 cm^{-1} which belong to Li_2CO_3, while some of them also have peaks of lithium alkyl carbonate ($ROCO_2Li$ 1650-1620 cm^{-1}). The latter peaks result from reaction of the lithium with unavoidably present trace O_2 and CO_2 in the glove box (to form Li_2CO_3) during sample preparation. $ROCO_2Li$ is formed due to a reaction of trace CO_2 with Li alkoxides.[9,10] In fact, these spectra are similar to those obtained during previous studies that proved that DN reacts with lithium to form Li alkoxides of the type $CH_3CH_2OCH_2OLi$ and others.[10] Hence, these studies prove that the T-amines, when present as additives, do not contribute to the build up of surface films formed on lithium in DN solutions. If they would indeed react with lithium when dissolved in DN, typical peaks of the aromatic rings would characterize spectra 4c and 4d (e.g. C-H stretching peaks above 3000 cm^{-1}).

X-ray microanalysis of Li samples stored in DN containing T-amines provided further proof that the amines' reaction with lithium is not important in these cases as no nitrogen peaks could be observed, while pronounced O and C peaks appear in the spectra. The reactivity and stability of the above mentioned T-amines under the various experimental conditions is summarized in Scheme 1.

The compatibility of the T-amines and their stabilizing effect is further proved by scanning electron microscopy of lithium electrodes cycled in $LiAsF_6$ and $LiClO_4$/DN-solutions. Figure 5 shows 4 micrographs obtained from Li electrodes charged-discharged in DN solutions. Micrographs 5a and 5c are related to DN/$LiAsF_6$ and DN/$LiClO_4$ solutions, respectively, with no additives, while Figs. 5b and 5d are related to DN/$LiAsF_6$/1000 ppm TBzA and DN/$LiClO_4$/1000 ppm TEA, respectively. These micrographs show that Li morphology during charge-discharge cycling is much smoother when T-amines are present in solutions (less dendrites are formed).

b. Li cycling efficiency in stabilized DN solution

Li cycling efficiency in DN/$LiAsF_6$ containing T-amines was measured (and tested) under a variety of conditions. Figure 6 shows the variation of Li cycling efficiency during half cell testing in DN/$LiAsF_6$ 1M solution stabilized with 100 ppm of tripropylamine as a function of D.O.D. and the number of cycles. The data related to DN-alkyl carbonate mixtures also appearing in this figure is discussed later. In these experiments, the electrodes were separated by a few mm of solution and the initial charge of the tested Li electrode (prepared by Li electrochemical deposition) was 5 C/cm^2 (C.D. was 1.5 mA/cm^2). As shown in Figure 6 average

147

cycling efficiencies around 94-98% could be obtained. In other series of experiments with DN/LiAsF$_6$/100 ppm TPrA solutions in the same type of cells,[7,10] the variation of Li cycling efficiency was measured as a function of the current density or of the amount of lithium in the tested electrodes. C.D. ranged from 1 to 10 mA/cm^2 and the charge in the working electrode ranged from 5 to 20 C/cm^2.

In this experimental series the average Li cycling efficiency ranged between 95-98%. The influence on the Li cycling efficiency of the different T-amines and their solution concentrations in similar half cells was also explored. TEA, TBzA and TBuA were tested as stabilizers in addition to TPrA. In all these experiments similar Li cycling efficiency around 95-98% was measured. These results further prove that the T amine does not interfere with the surface chemistry of lithium in solutions, as concluded above based on the spectral studies. In another series of experiments Li cycling efficiency was measured in DN/LiAsF$_6$ 1M solution containing various amounts of tripropylamine ranging from 100 ppm to 25% by volume. Even at TPrA content of 25%, Li cycling efficiency around 95%, was measured. This result is important since high content of T-amines in solution increases the safety in the use of DN solutions. In case of accidents, which lead to solvent evaporation and an exposure of lithium to the air, the T-amines, which are oily and have low vapor pressure, remain on the surface and prevent the dangerous reaction between lithium and air.

Several series of experiments were performed with the cell shown in Figure 1, in which the tests are closer to the real situation existing in the batteries as the electrodes are pressed together with a separator in between, allowing a very small volume of electrolyte solution in the cell. In these experiments the salt concentration ranged between 1 - 1.5M, the amount of charge in the tested electrode ranged between 5-10 C/cm^2 and the C.D.'s were 1-5 mA/cm^2. The average Li cycling efficiencies were between 95% and close to 100%.

The above experiments prove the superiority of stabilized LiAsF$_6$/DN for secondary Li systems compared to LiClO$_4$/DN that was also previously reported as an excellent electrolyte solution for rechargeable Li batteries.[1]

Figure 7 presents X-ray microanalysis data related to lithium electrodes treated in various DN solutions. The treatment included galvanostatic charge-discharge cycling (3-5 cycles, 1 mA/cm^2, 0.5 - 1 C/cm^2 per process). As demonstrated in this figure, spectra obtained from Li electrodes treated in DN/LiAsF$_6$ have pronounced fluorine and small arsenic peaks in addition to the expected C and O peaks. Spectra related to lithium treated in DN/LiClO$_4$ have chlorine peaks and the ratio of the O/C peaks is much higher than that obtained from samples treated in LiAsF$_6$ solutions. Hence, these results prove that LiAsF$_6$ in DN is reduced by lithium to LiF and possibly to insoluble arsenic compounds, and LiClO$_4$ is reduced to LiCl and Li$_2$O. (Li$_2$O formation, in addition to the chloride, explains the high O/C ratio). It can be concluded that the halides precipitating on lithium play an important role in determining cycling efficiency and morphology. This may be attributed either to their properties as Li$^+$ ionic conductors and/or their degree of interference with precipitation of homogeneous, ordered organic films.

c. Mixtures of DN and alkyl carbonates.

It was already reported that mixture of ethers (e.g. 2-MeTHF) and EC are very good electrolyte solutions for secondary Li batteries.[5] As mentioned in Scheme 1, PC or EC does not react with the T-amines in DN solutions. Figure 8 shows variation of Li cycling efficiency and conductivity obtained with DN/PC/LiAsF$_6$ solution as a function of % PC (by volume). For a comparison, Li cycling efficiency obtained with THF/PC/LiAsF$_6$ solutions is also presented. Cycling efficiency was measured in these cases with Li deposited on nickel (5 C/cm^2) WE which was charged-discharged 10 times (1.5 mA/cm^2, 25% DOD). As expected, addition of the polar PC to DN solution increases its conductivity. The effect of PC addition on Li cycling efficiency is more complicated. A maximum in average Li cycling efficiency was measured with mixtures containing 5-10% PC. In contrast, Li cycling efficiency deteriorates with addition of PC to THF, as also shown in Figure 8. As shown in Figure 9, somewhat similar behavior was observed with EC/DN/LiAsF$_6$ solutions (stabilized with T amine). An increase in Li C.E. (compared to DN/LiAsF$_6$) could be measured with solution containing 0.1-3% EC. Some other data related to Li cycling efficiency measured with EC/DN or PC/DN LiAsF$_6$ solutions close to the optimal ratios are presented graphically in Figure 6.

Morphology studies of Li electrodes treated in these solutions are in line with the variations in Li C.E. shown in Figure 8 and 9 as demonstrated in Figure 10, which presents SEM micrographs obtained from Li electrodes treated in DN,DN/PC and DN/EC solutions. As shown in micrographs 10 (a-h), Li electrodes treated in DN/LiAsF$_6$ containing low percentages of PC or EC (11b, 11c and 11g) are very smooth. As the amount of the alkyl carbonate in solution increases, the lithium surface becomes rougher (10d, 10e, 10f and 10h), which correlates well with the decrease in cycling efficiency as the percentage of PC or EC increases, as shown in Figures 8 and 9.

One would assume that addition of reactive cosolvents such as PC or EC to DN/LiAsF$_6$ affects Li cycling efficiency so pronouncedly due to the impact on LiAsF$_6$ reduction because of the competition of alkyl carbonate reduction on lithium to form stable passive films. However, X-ray microanalysis of lithium samples treated in PC, EC/DN mixture shows that LiAsF$_6$ reduction is not strongly affected by the presence of PC or EC in DN, as demonstrated in Figure 7. Hence, the presence of PC or EC probably affects mostly the Li surface chemistry related to solvent reactions. Figure 11 shows FTIR spectra obtained from nickel electrodes treated at Li deposition potential (0V Li/Li$^+$) in LiAsF$_6$/DN/EC and LiAsF$_6$/DN/PC. For a comparison, a spectrum obtained from nickel treated in LiAsF$_6$/THF/PC is also present. All the three spectra of Figure 11 are typical of lithium alkyl carbonates which are the PC and EC reaction product with lithium.[8] However, these spectra are different from each other in several wavenumber domains where DN or THF reduction products (Li alkoxides) absorb in the IR. There are pronounced differences in these spectra around 2900 - 2800 cm^{-1} (ν C-H), 1800 - 1350 (δ CH$_2$, CH$_3$) and around 1100 - 1000 (ν - CO). Hence, these spectral results prove that in spite of the reactivity of the alkyl carbonate, both solvents contribute to the build-up of surface films on lithium. This further proves that Li cycling efficiency depends mostly on a complicated surface chemistry which is determined by a delicate balance of several simultaneous film forming processes related to solvents and salt (and other active contaminants/additives, if present). Hence, at certain PC/DN or EC/DN ratios the surface film formed which contains both ROCO$_2$Li and ROLi compounds (resulting from DN reduction) allows improved, homogeneous Li-deposition and dissolution (as proved by the SEM studies) and, therefore, improved cycling efficiency. In contrast, the presence of alkoxides formed by THF reduction (mostly lithium butoxides) together with ROCO$_2$Li species on the lithium surfaces when PC/THF solutions are used, adversely affects Li cycling efficiency and morphology.

One would ask how, in spite of the superior reactivity of the alkyl carbonates toward lithium compared to ethers, the latter also strongly influences Li surface chemistry when ether/PC or EC mixtures are used. This may be explained as follows: Lithium does react predominantly with PC or EC to form a porous Li-alkyl carbonate film. It is possible that since PC or EC molecules are polar, their mobility to the Li surface through the pores is limited while the less polar ether molecules can better diffuse through the films and reach the Li surface and be reduced in the pores to form lithium alkoxides.

Conclusion

Tertiary amines stabilize DN-Li salt solutions even when present in trace amounts (100 ppm). The stabilized solutions have a shelf life of years. The T-amines disolved in DN do not interfere with Li surface chemistry in solutions. Stabilized DN-LiAsF$_6$ solutions are superior electrolyte solutions for secondary lithium batteries.

Li cycling efficiency of 96-98% was measured in a variety of tests at different conditions, part of which were very similar to those of real battery situations. Further improvement could be obtained by the use of PC or EC as cosolvents in DN/LiAsF$_6$/T amine solutions. In part of the tests Li cycling efficiencies of more than 98% could be measured at DN/PC and DN/EC ratios around 12:1 and 50-30:1, respectively. The use of alkyl carbonates as cosolvent in DN solutions has several advantages (in addition to the increase in C.E.):

1. Conductivity increases as EC or PC is added to DN/LiAsF$_6$.

2. Should leakage causing DN evaporation occur, dangerous contact between the lithium and air is avoided by the unvolatile PC or EC . A thin layer of alkylcarbonate remained on the lithium, protects it from vigorous reactions with water and oxygen, since a protective Li$_2$CO$_3$ layer is formed due to reaction of lithium, trace water and PC or EC.

The electrochemical windows of the stabilized DN/LiAsF$_6$ is 3.2-3.7 V (depending on the oxidation potential of the T-amine), making them compatibile with a variety of cathodes such as TiS$_2$, MoS$_2$, V$_2$O$_5$ and others. Since Li cycling efficiency depends mostly on a delicate balance of film forming reactions related to solvent salt and additives, it is believed that further optimization of these systems related to the amount of salt, cosolvent and stabilizer can lead to even better performances than those reported in this work.

Acknowledgement

Partial support for this work was obtained from the BSF Binational U.S.-Israel Science Foundation.

References

1. K.M. Abraham and S.B. Brummer in "Lithium Batteries" (Edited by J.P. Gabano), Chap. 14, Academic Press, N.Y. (1983).

2. G.H. Newmann in Proceedings of the Workshop on Lithium Non-Aqueous Battery Eelectrochemistry, P.V. 80-7, The Electrochemical Society, Inc., New Jersey, p. 143 (1980).

3. G.H. Newmann, R.W. Francis, L.H. Gaines and B.M.L. Rao, *J. Electrochem. Soc.* 127, 2025 (1980).

4. O. Youngman, P. Dan and D. Aurbach, *Electrochimica Acta.* 35, 639 (1990).

5. S. Tobishima, M. Arakawa, T. Hirai and J, Ymaki, *J. Power Sources* 26, 449 (1989)

6. D. Aurbach, *J. Electrochem. Soc.* 136, 1606 (1989).

7. D. Aurbach, Y. Gofer and Y. Langzam, *J. Electrochem. Soc.* 136, 3198 (1989).

8. D. Aurbach and H. Gottlieb, *Electrochimica Acta.* 34, 141 (1989).

9. D. Aurbach, M.L. Daroux, P. Faguy and E. Yeager, *J. Electroanal. Chem.* 297, 225 (1991).

10. O. Youngman, Y. Gofer, A. Meitav and D. Aurbach, *Electrochimica Acta.* 35, 625 (1990).

11. C.J. Pouchert, "The Aldrich Library of FTIR Spectra". 1st ed., Aldrich Chemical Company, Milwaukee, WI (1985).

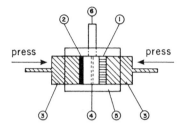

Figure 1. The small solution volume, compressed cell used for part of the Li cycling efficiency measurements: 1. Working electrode - nickel disk on which Li was deposited (5-20 C/cm²). 2. Counter electrode - Li disk. 3. Stainless steel current collectors. 4. Polypropylene separator and the solution volume. 5. Polyethylene cylinder cell body. 6. A glass tube for loading the cell with solution. After loading, the working and counter electrodes were pressed towards each other leaving a very small space between them for solution. The solution excess was removed through tube 6.

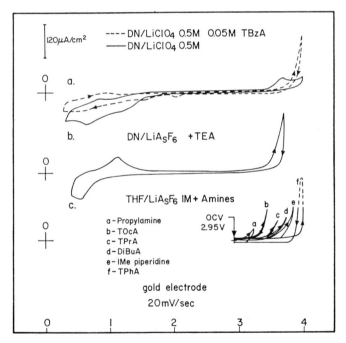

Figure 2. Typical cyclic voltammograms obtained with stabilized and unstabilized solutions. Gold electrode, 20 mv/sec. a. Solid line - unstabilized DN - $LiClO_4$ 0.5 M solution. Partial polymerization occurs at potentials above 3.2 V. Dashed line - same solution, stabilized with tribenzylamine. b. A typical voltammogram obtained with $DN/LiAsF_6$ - 1M, containing T amine (TEA, 1000 ppm in this case). c. Oxidation potentials of some tertiary amines (dissolved in $THF/LiAsF_6$ 1M, conc. ≈ 1000 ppm).

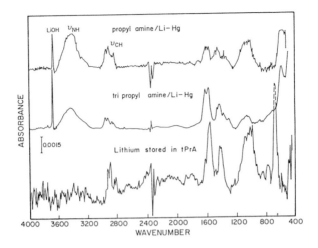

Figure 3. FTIR spectra obtained from reaction products of propyl amine and tripropylamine with Li/Hg amalgam (pelletized with KBr) and an FTIR spectrum obtained from a Li sample stored in tripropylamine for a few days for a comparison.

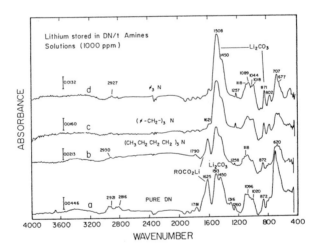

Figure 4. FTIR spectra obtained from Li surfaces stored in DN containing T-amines (1000 ppm) for a few weeks. a - pure DN for a reference, b - tributylamine, c - tribenzylamine, d - triphenylamine.

152

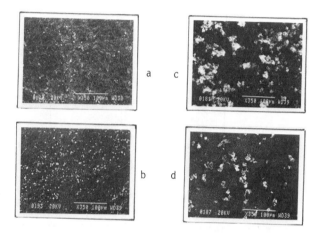

Figure 5. SEM micrographs obtained from Li electrodes treated in DN solutions (three charge-discharge cycles, 1.5 mA/cm^2, 0.5 C/cm^2 per process). A scale appears in each picture. a. DN/LiAsF$_6$ 1M solution, no additives. b. DN/LiAsF$_6$ 1M solution containing 1000 ppm tribenzylamine (TBzA). c. DN/LiClO$_4$ 1M solution, no additives. d. DN/LiClO$_4$ solution containing 1000 ppm triethylamine (TEA).

SCHEME 1

STABILITY AND REACTIVITY OF T-AMINES

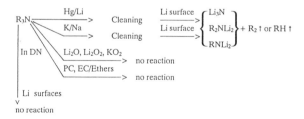

153

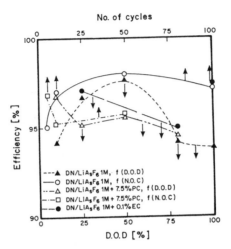

Figure 6. Variation in Li cycling efficiency obtained with DN/LiAsF$_6$ 1M, DN/LiAsF$_6$ 1M/PC 7.5% and DN/LiAsF$_6$ 1M/EC 0.1% solutions as a function of D.O.D. (10 charge-discharge cycles) and as a function of a number of cycles (D.O.D. = 25%). Working electrode was lithium on nickel (5 C/cm^2) and C.D. was 1.5 mA/cm^2. All the points in this figure are related to average efficiency at accuracy of ±1%.

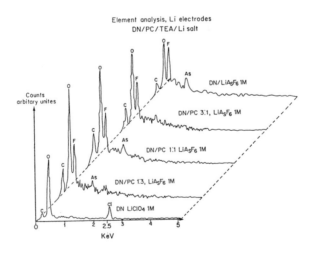

Figure 7. X-ray microanalysis data obtained from lithium electrodes treated in various DN solutions. (The treatment included 3 charge-discharge cycles, 1.5 mA/cm^2 0.5 C/cm^2 per process).

154

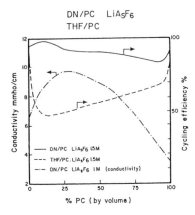

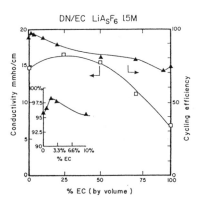

Figure 8. Variation in Li cycling efficiency and conductivity measured with DN/PC LiAsF$_6$ solutions as a function of % PC. Li cycling efficiency obtained with THF/PC/LiAsF$_6$ solutions is also presented for a comparison. The standard experiment included 10 charge-discharge cycles. (Li deposited on nickel or copper WE/ 5 C/cm^2, D.O.D. = 25%, 1.5 mA/cm^2).

Figure 9. Variation in Li cycling efficiency and conductivity measured with DN/EC/LiAsF$_6$ solutions as a function of %EC. The standard experiments are similar to those related to Figure 9.

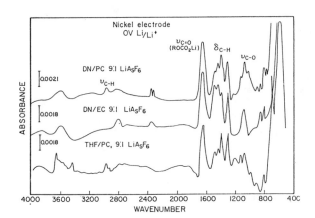

Figure 11. FTIR spectra (measured *ex situ*) obtained from nickel electrodes treated in DN/PC, DN/EC and THF/PC for a comparison. All solution contained 10% (by volume) alkyl carbonate and LiAsF$_6$ 1M. The treatment included a potentiostatic step from OCV ($\approx$2.5-3V, Li/Li$^+$) to 0.V (Li/Li$^+$). The electrodes were held at this potential for 15 minutes before washing (pure ether) and drying.

155

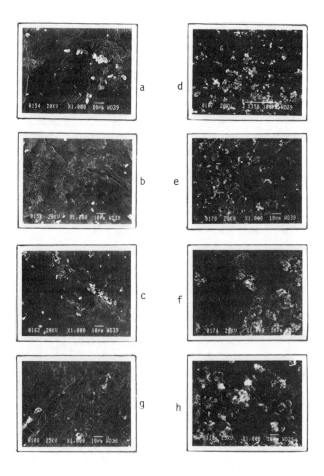

Figure 10. SEM micrographs obtained from Li electrodes treated in T amine stabilized DN/LiAsF$_6$ 1M solutions containing different amounts of PC or EC (3 charge-discharge cycles, 0.5 C/cm^2 per process, 1.5 mA/cm^2). A scale appears in each picture. a. 100% DN/100 ppm TEA. b. 1% PC/100 ppm TEA. c. 25% PC/100 ppm TEA. d. 50% PC/100 ppm TEA. e. 75% PC/100 ppm TEA. f. 1% EC/3000 ppm TBuA. g. 10% EC/3000 ppm TBuA.

THE ELECTROCHEMICAL BEHAVIOR OF METHYL FORMATE (MF) SOLUTIONS

Yair Ein Ely and Doron Aurbach
Department of Chemistry
Bar-Ilan University
Ramat Gan 52100, Israel

Abstract

The electrochemical behavior of methyl formate MF and MF/CO_2/Li salt solutions with active metals (lithium, calcium) and non-active metals (nickel, gold) was investigated. The salts included Ca, Li and tetrabutylammonium perchlorates and tetrafluoroborates, and $LiAsF_6$. Surface sensitive FTIR and X-ray microanalysis were used in order to characterize the various electrode surfaces, in addition to cyclic voltammetry. It was found that the metal formates (HCOOLi or $(HCOO)_2Ca$) are the major surface species formed both on the active and on the non-active electrodes at low potential and that they passivate them. When CO_2 and Li salt are present in MF solutions, films formed both on active and non-active electrodes at low potentials contain both Li formate and lithium carbonate.

Introduction

There is an increasing interest in methyl formate (MF) as a solvent or co-solvent for secondary lithium batteries. In a recent report related to $Li-LiCoO_2$ rechargeable cells, MF/CO_2 solution is described as a superior electrolyte for this system.[1] The conductivity of $MF/LiAsF_6$ solutions is around 30-50 mmho/cm and higher by an order of magnitude compared to propylene carbonate or ethereal solutions. Other reports related to Li-Vanadium oxide cells also mention MF solutions as a good electrolyte for these systems.[2] The compatibility of MF for secondary Li systems is surprising since esters are supposed to be very reactive with lithium. As previously reported,[1] the presence of CO_2 remarkably improves cycling efficiency of Li anodes in MF solutions. However, it is very unclear how the presence of CO_2 influences Li cycling efficiency in a solvent such as MF which is supposed to be very reactive with lithium. CO_2 reduction mechanisms in non-aqueous systems are not fully understood,[3] and it is not at all clear if CO_2 dissolved in MF is able to affect Li surface chemistry in solutions.

The present study aims to investigate the electrochemical behavior of MF and MF/CO_2 solutions with active (lithium, calcium) and non-active metal (gold, platinum, nickel) electrodes. A special effort was made to understand the surface chemistry of lithium, calcium and non-active metals at low potentials in MF solutions, using FTIR and X-ray microanalysis. In addition, the voltammetric behavior of MF and MF/CO_2 solutions was also studied using non-active metal electrodes and was compared to that of other polar aprotic systems previously studied.[4,5]

Experimental

MF (Aldrich, anhydrous) was distilled twice over P_2O_5 under argon at room temperature. $LiClO_4$ (Aldrich) was used after dehydration and drying in vacuum (150°C, three days); $LiAsF_6$ (Lithco) and $LiBF_4$ (Tomiyama) were used as received. FTIR measurements were carried out ex situ, using mirror-like lithium, calcium and non-active metal surfaces prepared as described elsewhere[4-6] (external reflectance mode). The experimental conditions, glove box operation, instrumentation, transfer systems and spectroscopic measurements were also described previously.[4-7]

For the study of MF/CO_2 solutions, mirror like lithium and non-active metals (gold, nickel) electrodes were studied in three electrode cells placed in aluminum pressure vessels. These were loaded with electrochemical cells (including MF solutions) in the glove box and were then pressurized with CO_2 (ultra high purity - Oxygen Center, Israel) to different pressures.

X-ray microanalysis of lithium and calcium samples was obtained with the eXL system (Link, England) attached to a JEOL 840-JSM electron microscope, using a transfer technique which was previously described.[7]

Results and discussion

a. MF solutions.

Figures 1 and 2 show FTIR spectra obtained from lithium and calcium electrodes (respectively) exposed to various MF environments. Spectra 1a and 2a were obtained from samples exposed to MF vapor, and the other spectra of figures 1 and 2 were obtained from samples stored in MF solution for a few hours as indicated. All the spectra of figures 1 and 2 have peaks around 2970, 2850, 2730 cm^{-1} (CH stretching), 1620 cm^{-1} (C=O stretching), 1380 cm^{-1} and 790 cm^{-1} (COO$^-$ bending) which are typical of formate anion (HCOO$^-$).[12] This proves that both lithium and calcium have similar surface chemistry in MF solutions producing metal formate as a major surface species. MF is probably reduced by the active metals to form a radical anion followed by a cleavage of the O-CH$_3$ bond to form metal formate and methyl radical which probably abstracts hydrogen to form methane or recombines to ethane. However, as proved in Figures 1 and 2, Li or calcium formates are not the only surface species formed in these systems. Other peaks around 1450, 1100-1000 and 700cm^{-1} appear in part of the spectra of Figures 1 and 2. Their appearance depends on the experimental condition and the presence of salts in solutions. It is interesting to note that while the spectra obtained from calcium surfaces exposed to MF vapor have typical CaCO$_3$ peaks (Fig. 2a, as indicated), spectra obtained from calcium surfaces identically prepared, stored in MF are mostly calcium formate spectra (Fig. 2b). CaCO$_3$ may be unavoidably present on calcium surfaces prepared in the glove box due to the presence of trace O$_2$ and CO$_2$ in glove box atmosphere.[11] Therefore, both CaCO$_3$ and Ca formate may be present on calcium surfaces exposed to MF vapor. These studies prove that in solutions, the initial CaCO$_3$ film is unstable and is readily substituted by the solvent reduction products.

Naturally, one would suspect Li or Ca methoxide to be formed as well, due to a cleavage of MF radical anion to the HCO radical and CH$_3$O$^-$. Fig. 3a shows FTIR spectra obtained from LiOCH$_3$ stored for a few days in MF, (dried and pelletized with KBr). The spectrum is identical to that of the starting alkoxide. No lithium formate was formed, which proves that Li methoxide is stable in MF. None of the LiOCH$_3$ peaks appear in the spectra of Fig. 1 and 2. Spectrum 3b, which was obtained from lithium surface exposed to methanol vapor followed by storage in MF, has both LiOCH$_3$ and LiOOCH peaks. Spectrum 3c obtained from lithium surface stored in MF/MeOH 0.01M solution has only the typical lithium formate peaks. These experiments obviously prove that methoxide is not formed in these systems and that if it was formed on lithium in MF it should be stable and detectable. Also, methanol, if present as a contaminant in MF, does not react with the active metal, probably because of the high reactivity of the solvent.

Another contaminant that may be unavoidably present in MF is trace water which may react with active metals to form insoluble hydroxides. Figures 4a and 4b show FTIR spectra of anhydrous LiOH and LiOH stored over MF respectively. The solid formed by the reaction of MF and LiOH was vacuum dried after storage. (The spectra of this product and LiOH were obtained pelletized with KBr). Spectrum 4b is typical of LiOOCH, which means that LiOH reacts completely with MF to form Li formate and methanol. Spectrum 4c obtained from lithium stored in MF/H$_2$O 0.01M solution is also typical of lithium formate. No LiOH is formed, as indicated by the absence of the typical 3675 cm^{-1} LiOH peak.

This proves that the surface chemistry of lithium in water-contaminated MF is similar to that of dry MF since LiOH, if formed, also readily reacts with MF to form Li formate. In contrast to that, as proven by spectrum 4d, Ca(OH)$_2$ if formed on calcium in wet MF only partly reacts with the solvent. The Ca formate formed probably passivates the hydroxide and in this case avoids a complete reaction. Among many polar aprotic solvents previously studied, including ethers, BL, propylene carbonate and ethylene carbonate, MF seems to be the most reactive one toward active metals. Li formate could be detected on Li surfaces exposed to MF vapor for a few minutes. Also, Ca formate could be detected on calcium surfaces stored for a few hours in MF. The high reactivity of MF toward lithium and calcium is supposed to affect possible salt reduction, however the FTIR spectra of Figs. 1 and 2 discussed above prove that the presence of salt anions such as AsF_6^- or ClO_4^- influence the surface chemistry of the active metal in solutions as well. The presence of BF_4^- salts affect MF reduction on lithium even more strongly (as will be shown later). However, X-ray microanalysis of lithium or calcium samples stored in ClO_4^-, AsF_6^- or BF_4^- salt solutions could not detect chlorine, arsenic boran or chlorine peaks.

b.   MF/CO_2 solutions.

The surface chemistry of lithium in MF/CO_2 solutions at different CO_2 pressures (1-6 atm) was investigated by *ex situ* FTIR spectroscopy of either mirror-like lithium surfaces or polished nickel foils. It was found that the surface films formed on nickel electrodes polarized in solutions to potentials close to that of Li bulk deposition are similar to those formed on lithium.

Figure 5 shows FTIR spectra obtained from lithium electrodes stored for 30 minutes in MF/CO_2 (3 atm) $LiClO_4$ and $LiAsF_6$ solutions (5a and 5b, respectively). Both spectra have lithium formate peaks (1640-1620, 1380 and 780cm^{-1}) and Li_2CO_3 peaks (1500-1450 and 880-865cm^{-1}.[5]) Spectrum 5a also has a pronounced band around 1100cm^{-1} which is attributed to a Cl-O bond (similar to that of $LiClO_4$). This may be related to Li perchlorate reduction products such as $LiClO_2$ or $LiClO_3$ that should be much less soluble in MF and therefore precipitate on the surface.

Spectrum 5b also has, in addition to Li formate and Li_2CO_3, peaks around 1700, 1300, 1150 and 900 cm^{-1}, which are discussed below. Figure 6 presents FTIR spectra obtained from nickel electrodes polarized to 0.5 and 0.05V (Li/Li$^+$) in MF/$LiAsF_6$ 1M solutions pressurized to 6 atm and 2 atm (6a - d, respectively). Spectra 6a-c mostly have Li formate and Li_2CO_3 peaks as indicated. The 700cm^{-1} peak appearing in these spectra is due to to As-F bond and may be attributed either to residual $LiAsF_6$ or to species such as Li_xAsF_y formed by $LiAsF_6$ reduction. The 3600cm^{-1} peaks appearing in spectra 6a-c should be attributed to OH species, probably due to hydration of the surface species.

Spectrum 6d is different from the other spectra of Fig. 6 having major peaks around 1300, 1150, 900 and 600 cm^{-1}. (These peaks also appear in spectrum 5b). Based on other studies with carbon electrodes which enabled extraction of surface species in an amount suitable to NMR and MS studies, this spectrum is attributed to species formed by complicated condensation of CO_2 and MF reduction products containing alkoxy groups.

These spectral studies obviously prove that the surface chemistry of lithium and noble metals at low potentials in MF/CO_2 based electrolytes depends both on potential and CO_2 pressure. As the CO_2 pressure is higher, the major surface species formed are Li formate and Li_2CO_3. CO_2 reduction in both aqueous and non-aqueous systems was, in recent years, intensively investigated. In aprotic solvents CO_2 reduction seems to be complicated and some reports list oxalate anion ($C_2O_4^{-2}$) carbonate anion (CO_3^{-2}) and CO as its possible products.[3]

The present study obviously proves that CO_2 dissolved in MF is reduced in the presence of Li salts both on lithium and noble metals to Li_2CO_3. The Li formate detected also is obviously formed by solvent reduction since in the absence of protic substances CO_2 can not be reduced to formate.

159

When the salt is $LiClO_4$, Li_2CO_3 may also be formed due to a reaction between Li_2O and CO_2. Since it is known that $LiClO_4$ is reduced in aprotic media to $LiCl$,[8] Li_2O is an obvious co-product. This oxide readily reacts with CO_2 to form lithium carbonate.

Hence, the improvement of Li cycling efficiency obtained in MF based electrolyte due to addition of CO_2[1] should be attributed to a better passivation of the lithium electrodes due to Li_2CO_3 formation. It should be noted that enhancement of Li_2CO_3 formation over organic surface species on Li electrodes in PC based electrolyte solutions (for instance, by an addition of trace H_2O to PC solutions) also increases Li cycling efficiency in these systems.[7]

However, the best MF based electrolyte system for rechargeable Li batteries was reported to be $MF/LiAsF_6/LiBF_4/CO_2$ solution.[1] Therefore, special attention was given to this system. Figure 7 shows FTIR spectra obtained from nickel electrodes polarized in $MF/LiBF_4/CO_2$ and $MF/LiBF_4/LiAsF_6/CO_2$ solutions to low potentials (0.05 V, Li/Li$^+$). Spectra obtained from samples treated in similar solutions containing no CO_2 are also present for a comparison (spectra obtained from lithium samples stored in these solutions were quite similar but of a worse quality). Comparing these spectra to those of Figs. 1-6 shows that the presence of $LiBF_4$ in solutions affects the surface chemistry remarkably. Except for spectrum 7c obtained from $LiAsF_6/LiBF_4$ solution, none of the spectra have the clear, high resolution lithium formate and/or Li_2CO_3 peaks appearing in the previous figures. The pronounced peak around 1100cm^{-1} appearing in spectra 7a and 7d may be attributed to B-F bond, as $LiBF_4$ spectrum has a similar peak. The 1620cm^{-1} peak appearing in all the spectra of Fig. 7 probably proves that Li formate is formed but it is obviously not the major surface species.

Hence, one would conclude that the presence of BF_4^- changes the surface chemistry lithium electrodes, perhaps due to BF_4^- reduction to form insoluble products which block the surface and suppress the aforementioned surface reactions in solutions containing no $LiBF_4$. Hence, a new combination of surface films is obtained which occasionally leads to better passivation of the lithium electrodes and probably a more homogeneous Li deposition and dissolution, which may explain the high performance of this electrolyte system.[1]

c.   The voltammetric behavior of MF solutions.

Figures 8 and 9 show typical first cycle voltammograms obtained with gold electrodes and $MF/LiClO_4$ and $MF/LiAsF_6$ solutions, respectively (dashed lines). The effect of CO_2 atmosphere (3 atm) is also shown in these figures (solid lines). In both cases, the currents measured during the cathodic sweep at low potentials are at least two fold higher than those measured with other polar aprotic solvents containing Li salts in similar experiments.[4,5] However, these currents become much smaller in consecutive sweeps, which means that the electrodes become passivated due to precipitation of reaction products. The typical peak of lithium UPD on gold and its related anodic stripping peak which characterizes voltammograms of other non-aqueous systems containing Li salts are scarcely visible in the voltammograms of Figures 8 and 9. The peak around 1.25V appearing in both figures is attributed to water reaction, as addition of water to the solutions leads to its increase.

Massive solvent oxidation occurs at potentials above 5V and the oxidation products do not passivate the electrodes. The anodic peaks between 4 and 5V are related to an oxidation of surface films formed during the cathodic sweep as proven by the spectroscopic results to be presented below. Figure 10 shows FTIR spectra obtained ex situ from gold electrodes treated at different potentials in $MF/LiAsF_6$ solutions. In each experiment the potential was swept from OCV ($\approx$3V, Li/Li$^+$) to a predetermined low potential (followed by washing with pure MF and drying prior to the measurement). Spectrum a related to 1.6V is attributed to either traces of

160

partly hydrated salt or salt reduction products as the peak around $720cm^{-1}$ is related to As-F bond and the peaks around 3600-3500 and $1630cm^{-1}$ are typical of trace water hydrating Li salts.

However, as the potential becomes lower, typical Li formate peaks characterize the spectra (Figs. 10b and c), and their height increases as potential decreases.

As proven by spectrum 10d, measured after the potential was swept from 0CV to 0.1V and then to 4.5V, the lithium formate surface species are oxidized around 4.5V. Similar spectra results were obtained from electrodes treated in $LiClO_4/MF$.

Figure 11 shows voltammograms measured with gold electrodes treated in $MF/Ca(ClO_4)_2$ solution. In this case the reference electrode was a calcium foil in solution, which is about 0.7V in the Li/Li^+ scale. FTIR measurements of electrodes treated in this system at different potentials seems to indicate that the cathodic currents at potentials below 1V (Ca/Ca^{++}) are related to solvent reduction to calcium formate that precipitates on the electrode surface and partly passivates it. Hence, the surface chemistry of noble metals electrodes at low potentials in MF solutions of Ca and Li salts is quite similar to that of lithium or calcium in these systems. The solvent is reduced at potentials below 1.5V (Li/Li^+) to Ca or Li formate which passivate the electrodes due to precipitation. Lithium can be easily deposited through the Li formate films. In contrast, calcium can not be deposited, probably due to the poor mobility of the bivalent Ca^{++} through the surface films. As shown in Figures 8 and 9, the presence of CO_2 changes the voltammetric behavior. In the case of $LiClO_4$ solutions, lower currents are measured at low potentials (compared to solutions under argon atmosphere). When the salt is $LiAsF_6$, a pronounced irreversible reduction wave was measured at potentials below 0.5V.

Figures 12 and 13 present FTIR spectra obtained from nickel electrodes treated at various potentials in MF/CO_2 (3 atm) solutions of $LiAsF_6$ and $LiClO_4$, respectively. The spectra of Figure 12 prove that MF is reduced on non-active metals in the presence of Li salts around 1.5V, but CO_2 reduction requires lower potentials. As the potential is lowered, Li_2CO_3 is formed (in addition to Li formate) and becomes the major surface species. However, at potentials close to that of Li deposition potentials (Figure 13c) new products, whose structure will be discussed elsewhere, are formed. The reduction wave in Figure 9 is probably due to this process. The surface chemistry of the electrodes treated in $LiClO_4/MF/CO_2$ solutions is complicated. Spectrum 13a related to 1.5V has both Li formate and Li_2CO_3 peaks. However, since spectrum 12a related to $LiAsF_6$ solution has no Li_2CO_3 peaks, it is obvious that the source of the Li carbonate formed on the electrode in $MF/CO_2/LiClO_4$ around 1.5V is not CO_2 reduction but rather a secondary reaction. Hence, these results seem to prove that in spite of the competition of solvent reduction, $LiClO_4$ is also reduced around 1.5V. An obvious product is Li_2O (in addition to chlorides) which readily reacts with CO_2 to form Li_2CO_3.

Figure 13b related to 1.V also has, in addition to Li formate and Li_2CO_3 peaks, $ROCO_2Li$ peaks (1320, 820 and the shoulders around 1650 and 1090 cm^{-1}). These species could be formed due to reaction between alkoxy and CO_2.[4]

Figure 13c related to 0.05V shows that in contrast with $LiAsF_6/MF/CO_2$, the relative amount of Li_2CO_3 does not increase as potential decreases (compare with Figures 12b-c).

Hence, the passivation of the electrodes at low potentials in $LiClO_4/MF$ solutions due to the presence of CO_2 as indicated in Figure 8 may be explained by the above secondary reactions of CO_2 with surface species formed by salt and solvent reduction, which suppress direct CO_2 reduction, probably due to electrode blocking by their products.

Conclusion

Methyl formate is one of the most reactive polar aprotic solvents toward active metals. It is reduced on active metals(such as lithium and calcium) to metal formate as a major product which precipitates on the metal surface and passivates it. MF reduction on noble metals in the presence of Li or Ca salts occurs at potentials below 1.5V (Li scale) to also form metal formate, which forms surface films and passivates the electrodes.

The presence of two expected contaminants - water and methanol in solutions at trace amounts - does not affect the surface chemistry of active metals or noble metals at low potentials in MF since traces of methanol do not seem to react on the electrodes, and metal hydroxide formed by water reduction reacts with MF to form metal formate.

The compatibility of $MF/CO_2/Li$ salt solutions for secondary Li system reported, which is surprising in light of the high reactivity of MF, is due to the effect of CO_2 on the surface chemistry. CO_2 in MF is reduced on lithium and on non-active metals at low potential to Li_2CO_3, therefore passive films formed on lithium or noble metals at low potentials in MF/CO_2 contain both Li formate and Li carbonate.

However, when the salt is $LiAsF_6$ and at low CO_2 pressures, MF/CO_2 solutions are reduced on lithium to a more complicated species, probably polymeric, whose identification has not yet been completed.

Acknowledgement

Partial support for this work was obtained from BSF Binational US-Israel Science Foundation.

References

1.   E. Plichta, S. Slane, M. Uchiyama, M. Salomon, D. Chua, W.B. Ebner and H.W. Lin, *J. Electrochem. Soc.* **136**, 1865 (1989).

2.   M. Uchiyama, S. Slane, E. Plichta and M. Salomon, *J. Power Sources,* **20**, 279 (1987).

3.   I. Taniguchi in "Modern Aspects of Electrochemistry", Vol. 20. J.O'M Backris, R.E. White and B.E. Conway eds., Plenum Press, N.Y. and London (1991), pp. 327-344.

4.   D. Aurbach, M.L. Daroux, P. Faguy and E. Yeager, *J. Electroanal. Chem.* **297**, 225 (1991).

5.   D. Aurbach and H. E. Gottlieb, *Electrochimica Acta*, **34**, 141 (1989).

6.   D. Aurbach, *J. Electrochem. Soc.* **136**, 1606 (1989).

7.   D. Aurbach, Y. Gofer and Y. Langzam, *Ibid.*, **136**, 3198 (1989)

8.   M. Garreau, J. Thevenin and D. Warin, *Prog. Batt. Solar Cells* **2**, 54 (1979); M. Froment, M. Garreau, J. Thevenin and D. Warin, *J. Microsc. Spectrosc. Electron.* **4**, 111 483 (1979).

162

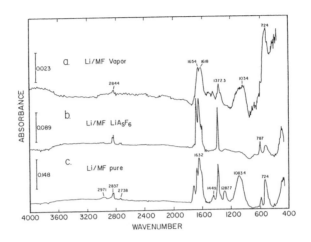

Figure 1. FTIR spectra obtained *ex situ* from lithium surfaces stored in MF and MF solutions. a. exposure to MF vapor for an hour; b. MF/LiAsF$_6$ 0.5M (a few hours); c. pure MF (for a few hours).

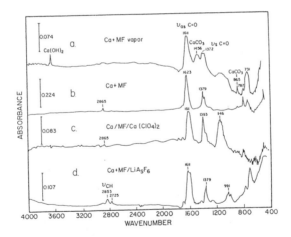

Figure 2. FTIR spectra obtained *ex situ* from calcium surfaces stored in MF and MF solutions. a. exposure to MF vapor (a few hours); b. MF/Ca(ClO$_4$)$_2$ 0.5M (a few days); d. MF/LiAsF$_6$ 0.5M (a few days).

163

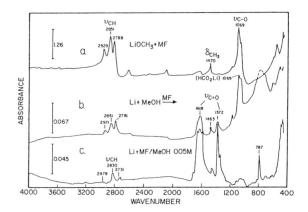

Figure 3. a. FTIR spectrum of LiOCH$_3$ stored for a few days in MF (pelletized with KBr after drying); b. FTIR spectrum obtained from lithium surface exposed to methanol vapor (a few minutes) followed by storage in MF (a few hours); c) FTIR spectrum obtained from Li surface stored for a few hours in MF/MeOH 0.05M solution.

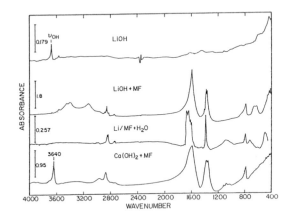

Figure 4. a. FTIR spectrum of KBr pelletized LiOH; b. FTIR spectrum obtained from LiOH stored for a few days in MF (pelletized with KBr after drying); c) FTIR spectrum obtained from lithium surface stored in wet MF (0.01M H$_2$O) for a few days; d) FTIR spectrum of Ca(OH)$_2$ stored for a few days in MF (KBr pelletized after drying).

164

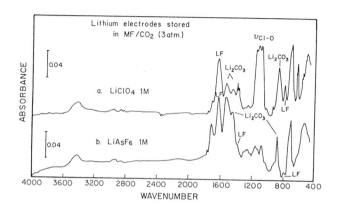

Figure 5. Typical FTIR spectra spectra obtained from lithium surfaces stored for half-an-hour in MF/CO$_2$ (3 atm) solutions containing Li salts. a. LiClO$_4$ 1M. b. LiAsF$_6$ 1M.

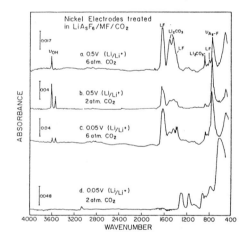

Figure 6. FTIR spectra obtained *ex situ* from nickel electrodes polarized to 0.5 and 0.05V (Li/Li$^+$) in LiAsF$_6$ 1M/MF/CO$_2$ solutions. a. 0.5V, 6 atm CO$_2$. b. 0.5V 2 atm. CO$_2$. c. 0.05V, 6 atm CO$_2$. d. 0.05V, 2 atm CO$_2$.

165

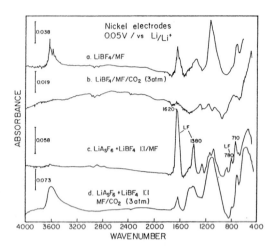

Figure 7. FTIR spectra obtained *ex situ* from nickel electrodes polarized to 0.05V (Li/Li$^+$) in various LiBF$_4$/MF solutions. a. LiBF$_4$ 1M/MF. b. LiBF$_4$ 1M/MF/CO$_2$ (3 atm). c. LiAsF$_6$ 0.5M/LiBF$_4$ 0.5/MF. d. LiAsF$_6$ 0.5M/LiBF$_4$ 0.5/MF/CO$_2$ (3 atm).

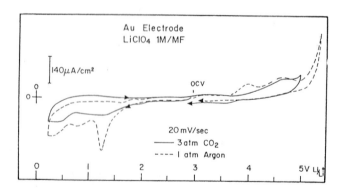

Figure 8. Typical first cycle voltammograms obtained with gold electrodes and MF/LiClO$_4$ solutions (20 mV/sec). Dashed line - argon atmosphere, solid line - CO$_2$ atmosphere (3 atm).

166

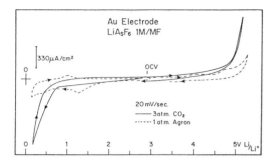

Figure 9. Typical first cycle voltammograms obtained with gold electrodes and MF/LiAsF₆ solutions (20 mV/sec). Dashed line - argon atmosphere, solid line - CO₂ atmosphere (3 atm).

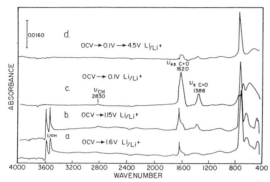

Figure 10. FTIR spectra obtained *ex situ* from gold electrodes treated in MF/LiAsF₆ 0.5M solution. The potential was swept from OCV to predetermined potentials as indicated in spectra a-d in the figure.

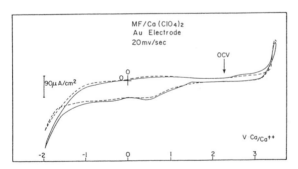

Figure 11. Typical voltammograms obtained with MF/Ca(ClO₄)₂ 0.5M solutions and gold electrodes (20 mv/sec). The solid line and the dashed line are related to first and second cycles, respectively

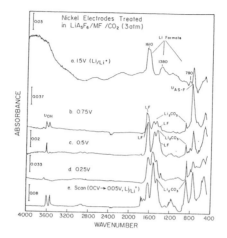

Figure 12. FTIR spectra obtained *ex situ* from nickel electrodes polarized in $LiAsF_6$ 1M/MF/CO_2 (3 atm) solutions. a - d. The potential was stepped from 0CV (~3V, Li/Li$^+$) to 1.5V, 0.75V, 0.5V and 0.25V, respectively. e. The potential was scanned from 0CV to 0.05V (Li/Li$^+$).

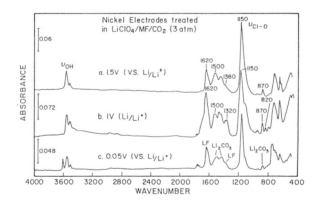

Figure 13. FTIR spectra obtained *ex situ* from nickel electrodes polarized in $LiClO_4$ 1M/MF/CO_2 (3 atm) solutions. a - d. The potential was stepped from 0CV (~3V, Li/Li$^+$) to a certain low potential followed by washing (pure MF) and drying. a. 1.5V; b. 1V; c. 0.05V.

DEVELOPMENT OF A LITHIUM SECONDARY BATTERY FOR A LOAD CONDITIONER --- EFFECTS OF ORGANIC ELECTROLYTES ON $Li_{1+x}V_3O_8$ CATHODE PERFORMANCE FOR SECONDARY LITHIUM CELLS ---

K. Takei, N. Terada, K. Ishihara, T. Iwahori and T. Tanaka
Central Research Institute of
Electric Power Industry (CRIEPI)
11-1 Iwato-kita 2-chome, Komae-shi, Tokyo 201, Japan

H. Mishima and K. Takeuchi
Yuasa Battery Co., Ltd.
6-6 Jyosai-cho, Takatsuki-shi, Osaka 569, Japan

$Li_{1+x}V_3O_8$ in an EC-based electrolyte showed better electrochemical behaviour with regard to reversibility than in a PC-based electrolyte for secondary lithium cells. This result suggests that cathode performance is related to the electrolyte decomposition by the reduction product of $Li_{1+x}V_3O_8$ and to the solubilites of $Li_{1+x}V_3O_8$ in electrolytes. Prismatic prototype cells using a $Li/Li_{1+x}V_3O_8$ system were fabricated and tested. These cells had energy densities of 114 Wh/kg and 211 Wh/l and showed energetic efficiency of 95 %.

Introduction

A load conditioner is a small-scale battery energy storage system that will be installed on the customer side[1]. Through the use of lithium secondary batteries, which are higher in energy density than conventional lead-acid batteries, the load conditioner is expected to be made into a lightweight and compact system. We have carried out the development of a large-scale lithium battery for load conditioner use.

Bronze $Li_{1+x}V_3O_8$ is one of the promising cathode materials for a high energy density secondary lithium cell because its lithiated solid solution is formed in a wide composition range[2]. However, poor electrochemical stability was observed on a $Li/Li_{1+x}V_3O_8$ cell in a PC/DME-based electrolyte. This work was undertaken to improve the electrochemical stability and cycling performance of $Li/Li_{1+x}V_3O_8$ cells in some kinds

169

of typically known electrolytes. Based on these results, prismatic prototype cells of 10 Ah -class capacity were fabricated and tested.

Experimental

Preparation of Li$_{1+x}$V$_3$O$_8$ and electrochemical measurements of 10 mAh test cell

Li$_{1+x}$V$_3$O$_8$ used in this study was prepared by melting Li$_2$CO$_3$ and V$_2$O$_5$ in vessels at 680 ^OC, as suggested by Wadsley[3]. The crystalline structure and composition of Li$_{1+x}$V$_3$O$_8$ thus prepared were checked by the X-ray diffraction method andatomic absorption spectroscopy. In this study we used Li$_{1+x}$V$_3$O$_8$ synthesized under the same conditions.

A cathode mixture sheet was obtained by mixing Li$_{1+x}$V$_3$O$_8$, acetylene black and Teflon and then rolling. The electrode was prepared by pressing the sheet on Ti foil. Test cells of 10 mAh-class capacity were hermetically sealed in a steel can in dry air. Li foil was used as both reference and counter electrodes. Cathode performance at 25 ^OC was examined by galvanostatic charge/discharge and cyclic voltammetry.

Preparation of electrolytes

The electrolytes for testing cathode performance were prepared from a mix of the high and low viscosity solvents (1 : 1). The compositions of the electrolytes are shown in Table 1[4]. EC (ethylene carbonate) and PC (propylene carbonate) were used as the high viscosity solvents. DME (1,2-dimethoxyethane) and 2MeTHF (2-methyltetrahydrofran) were used as the low viscosity solvents. 1 M LiClO$_4$ and 1.5 M LiAsF$_6$ were used as the solute. Each electrolyte was contained under 20 ppm H$_2$O.

Solubility of Li$_{1+x}$V$_3$O$_8$ powder

Li$_{1+x}$V$_3$O$_8$powder was immersed and stirred in various electrolyte components and conposition as shown in Table 1 (the ratio of 0.4 g to 100 ml) in Ar atomosphere. After filtering with Teflon filter paper (pore diameter : 0.45 um) and diluting with 0.1 M HNO$_3$, the vanadium ion concentration in the solutions was measured by inductively coupled plasma spectroscopy (ICP).

Fabrication of 10 Ah-class prototype cell

Prismatic prototype cells using a $Li/Li_{1+x}V_3O_8$ system were fabricated and tested.these cells had 1000-fold up-scaled capacities of the test cell based on cathode material. Cell performance was examined by galvanostatic charge/discharge. The temperature on the side face of the cell was measured in cycling tests.

Results and Discussion

Cathode performance by 10 mAh test cell

The potential curve and its differential values of prepared $Li_{1+x}V_3O_8$ are shown in Figure 1 as a function of Y in $Li_{1+x+y}V_3O_8$. The potential curve has at least 4 plateaus and it shows that the reaction of lithium intercalation into $Li_{1+x}V_3O_8$ has 4 different energy states. This bronz has a wide lithiation range (y > 3) and is thus a promising cathode material for secondary lithium cells.

Typical cyclic voltammograms of the initial 5 cycles with a low sweep rate of 3 mV/min in various electrolytes as shown in Table 1 are shown in Figure 2. In [B] electrolyte, an extra peak of anodic current of 3.1 V was observed in the first cycle. This peak was irreversible and the cyclability was not very good. The cyclic voltammograms in [D] and [F] electrolytes without PC showed good reversibility. In [C] and [E] electrolytes whose solutions were composed of PC/2MeTHF, the cyclabilities of voltammograms were not very good. The peak of anodic current shifted to higher potential while the peak of cathodic current shifted to lower potential with the number of cycles. In [G] electrolytes, anodic current rapidly increased at over 3.1 V; this was irreversible. Similar results were obtained in [A] electrolyte with some cells. Consequently, these results were not caused by short circuit between the cathode and anode. This behavior would be associated with the reaction of electrolyte decomposition.

Figure 3 shows cyclic voltammograms in [G] electrolyte in the case of different cathodic limit potentials of 2.6, 2.5 and 2.4 V. The irreversible oxidation reaction occurred in the case of the reduction of under 2.5 V. From these results, the oxidation reaction of the electrolyte decomposition was considered to be related to the reduction product of $Li_{1+x}V_3O_8$; i.e., $Li_{1+x}V_3O_8$ reduced sufficiently might act as a catalyst for the electrolyte oxidation.

Typical cycle performances of $Li_{1+x}V_3O_8$ in various electrolytes are shown in Figure 4. These results on cyclability agree with the results obtained by cyclic voltammetry. In the $LiClO_4$ group, EC/DME showed the best cycle performance. The initial specific capacity was about 240 mAh/g and it was almost constant with cycles. The coulombic efficiency was stable and near 100 %. In the $LiAsF_6$ group, EC/2MeTHF showed the best cycle performance but the specific capacity slightly decreased with the number of cycles. The cycle performance of the specific capacity in EC/PC was almost the same as that in EC/2MeTHF. But the coulombic efficiency was unstable and varied between 80 and 110 %. This was considered to cause the electrolyte decomposition. In [C] and [E] electrolytes, the specific capacity faded monotonously. PC/2MeTHF may be uncongenial with $Li_{1+x}V_3O_8$. In [B] electrolyte, the capacity decreased with the number of cycles but the fading was not constant.

Figure 5 shows the typical charge/discharge curves of 2nd, 5th and 10th cycles of $Li_{1+x}V_3O_8$ in various electrolytes. The curve of the 2nd cycle in [B] electrolyte had a shoulder at 3.1 V in charge, which corresponded to the anodic peak current of the cyclic voltammogram shown in Figure 2. In [G] electrolyte, a shoulder was observed in each cycle but was much sharper than that in [B] electrolyte. The reaction mechanism of these behaviours is not clear at present. We considered that the unstable cycle performance of the cathode was caused not only by the electrolyte decomposition mentioned above, but also by the solubility of $Li_{1+x}V_3O_8$ in electrolytes, because the decrease of specific capacity in [B] electrolyte was not constant.

Hence, the solubility of $Li_{1+x}V_3O_8$ in the components of [B] electrolyte was examined. Immersion time dependence on vanadium concentration in various electrolytes is shown in Figure 6. In each component of [B] electrolyte containing PC, the solubility of $Li_{1+x}V_3O_8$ increased with immersion time. On the other hand, in [E] electrolyte, soluble vanadium was hardly detected for over 200 h of immersion time.

Futhermore, Figure 7 shows the effect of the solute upon the solubility. Compared with the solubility in two kinds of the same solvent, $LiClO_4$ was considered to influence the solubility of $Li_{1+x}V_3O_8$. As a result of testing many kinds of electrolytes, $Li_{1+x}V_3O_8$ was found to be more soluble in electrolytes containing $LiClO_4$ than in electrolytes containing $LiAsF_6$. But the reproducibility of soluble quantities was not very good because the solubility was affected by very small amounts of water. At present, the relationship between the solubility of $Li_{1+x}V_3O_8$ and cathode cyclability is not yet established.

In summary, the following points are clarified.

(1) $Li_{1+x}V_3O_8$ showed good cycling performance in the electrolyte without PC.

(2) The cause of poor cycle performance was considered to be electrolyte decomposition and the solubility of $Li_{1+x}V_3O_8$.

(3) The results of cyclic voltammetry suggested that the oxidation reaction of the electrolyte decomposition was related to the reduction product of $Li_{1+x}V_3O_8$.

(4) $LiClO_4$ influenced the solubility of $Li_{1+x}V_3O_8$, but the relationship between the solubility of $Li_{1+x}V_3O_8$ and cathode performance was not clarified

Cell performance of 10 Ah prototype cell

Based on these results, we fabricated and tested prismatic prototype cells of a 10 Ah-class capacity. The specifications and initial characteristics of this prototype cell are shown in Table 2. Charge/discharge currents were 1.25 A / 1.25 A, respectively. This corresponded to the rate of about 0.1 C / 0.1 C. Charge/discharge preset voltge limits were 3.1 V / 2.0 V, respectively. Experimental data were the average from the 2nd to 10th cycles. These cells had gravimetric and volumetric energy densities of 114 Wh/kg and 211 Wh/l, respectively, which were about three times larger than those of conventional lead-acid batteries. Initial cycling data showed high energetic efficiency of 95%

The curves of charge/discharge of prototype cell are almost the same as those of the testing cells of a 10 mAh-class capacity. The temperature on the side face of the cell was increasing only about 1 ºC higher than the surroundings at the end of discharge. We succeeded in improving the high performance secondary lithium cells. But such problems as of cycle life, safety and so on, remains. We will continue the development of secondary lithium cells for practical use.

References

(1) K. Ishihara, Y. Nitta, R. Ishikawa and T. Tanaka; Proc. Sym. on Stationary Energy Storage: Load Leveling and Remote Application, 1987, The Electrochemical Society, Pennington, NJ, Vol.88-11, p.30 (1988).

(2) G. Pistoia, M. Pasquali, M. Tocci, R. V. Moshtev and V. Manev; *J.Electrochem.Soc.*, Vol.132, No.2, p.281 (1985).

(3) A. D. Wadsley; *Acta Cryst.*, Vol.10, p.261 (1957).

(4) G. Pistoia, M. Pasquali, Y. Geronov, V. Manev and R. V. Moshtev; *J.Power Sources,* Vol.27, p.35, (1989).

Table 1 Electrolyte compositions for electrochemical tests of $Li_{1+x}V_3O_8$.

Electrolyte	Solute	Solvent	
		High viscosity	Low viscosity
A	1M LiClO4	PC	
B	1M LiClO4	PC	DME
C	1M LiClO4	PC	2MeTHF
D	1M LiClO4	EC	DME
E	1.5M LiAsF6	PC	2MeTHF
F	1.5M LiAsF6	EC	2MeTHF
G	1.5M LiAsF6	EC+PC	

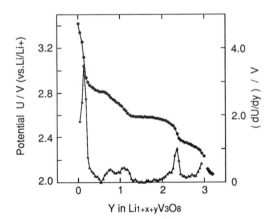

Figure 1 Final rest potentials (●) in current pulse method (0.2 mA/cm2, 1 h on, 4 h rest) as functions of y in $Li_{1+x+y}V_3O_8$ and their differential coulometric titration curves (▲).

174

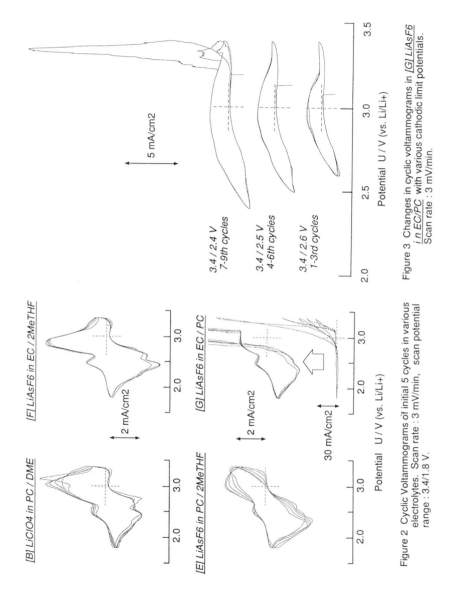

[B] LiClO4 in PC / DME

[E] LiAsF6 in PC / 2MeTHF

[F] LiAsF6 in EC / 2MeTHF

[G] LiAsF6 in EC / PC

2 mA/cm2

2 mA/cm2

30 mA/cm2

5 mA/cm2

3.4 / 2.4 V
7-9th cycles

3.4 / 2.5 V
4-6th cycles

3.4 / 2.6 V
1-3rd cycles

Potential U / V (vs. Li/Li+)

Figure 2 Cyclic Voltammograms of initial 5 cycles in various
electrolytes. Scan rate : 3 mV/min, scan potential
range : 3.4/1.8 V.

Figure 3 Changes in cyclic voltammograms in [G] LiAsF6
in EC/PC with various cathodic limit potentials.
Scan rate : 3 mV/min.

175

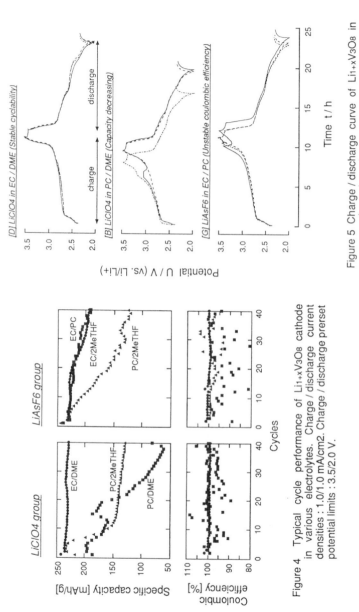

Figure 4 Typical cycle performance of $Li_{1+x}V_3O_8$ cathode in various electrolytes. Charge / discharge current densities : 1.0/1.0 mA/cm2. Charge / discharge preset potential limits : 3.5/2.0 V.

Figure 5 Charge / discharge curve of $Li_{1+x}V_3O_8$ in various electrolytes.
——: 2nd cycle - - -: 5th cycle ---: 10th cycle.

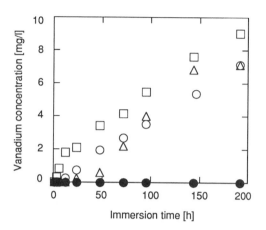

Figure 6   Vanadium concentration in various solutions
after $Li_{1+x}V_3O_8$ powder immersion.
□ : PC,   △ : [A] $LiClO_4$ in PC
○ : [B] $LiClO_4$ in PC / DME
● : [E] $LiAsF_6$ in EC / 2MeTHF

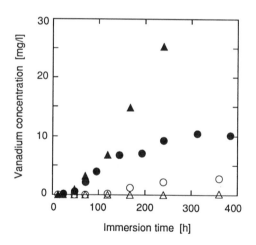

Figure 7 Comparison of solubility of $Li_{1+x}V_3O_8$
in $LiClO_4$- and $LiAsF_6$-based electrolytes.
● : [A] $LiClO_4$ in PC
▲ : [C] $LiClO_4$ in PC / 2MeTHF
○ : $LiAsF_6$ in PC
△ : [E] $LiAsF_6$ in PC / 2MeTHF

Table 2   Specifications and intial characteristics of $Li/Li_{1+x}V_3O_8$ prototype cell.

Specifications	Experimental data Average of 2-10th cycle		
		Charge	Discharge
Size 21w x 100l x 75(85)h mm	Capacity	12.7 Ah	13.0 Ah
Volume 157.5(178.5) cm3	Energy	35.0 Wh	33.3 Wh
Weight 291.7 g	Average voltage	2.75 V	2.54 V
( ) : contained terminal			
Nominal 10 Ah	Energy density	114 Wh/kg	
capacity 25 Wh		211 Wh/l	
	Ah efficiency	102 % ($\pm$1 %)	
	Wh efficiency	95.1 %	

INTERFACE BETWEEN SOLID POLYMER ELECTROLYTE AND LITHIUM ANODE

Zen-ichiro TAKEHARA, Zempachi OGUMI, Yoshiharu UCHIMOTO
and Eishi ENDO
Department of Industrial Chemistry, Faculty of Engineering,
Kyoto University,
Sakyo-ku, Kyoto 606-01, Japan

ABSTRACT

Thin-film solid-state lithium batteries of total thickness less than 20 μm were fabricated using a thin film of TiS_2 prepared by chemical vapor deposition (CVD) as the cathode active material, a thin film of solid polymer prepared by plasma polymerization as the electrolyte, and a thin film of lithium deposited by thermal evaporation as the anode. The solid polymer electrolyte film was formed by complexation of plasma-polymerized tris(2-methoxyethoxy)vinylsilane with lithium perchlorate. The relatively poor discharge performance of the thin film batteries was ascribed to the interfacial resistance between the electrolyte and the lithium anode. Thus, the effects of the interfaces between the solid polymer electrolyte and the lithium anode on the battery's discharge performance are discussed. FT-IR measurements showed that the interfacial resistance between the lithium electrode and the solid polymer electrolyte film likely was due to the formation of a resistive layer consisting of a mixture of lithium alkoxides and lithium alkylsilanolates.

1. INTRODUCTION

Recently, a significant amount of work has been devoted to solid-state lithium batteries utilizing solid polymer electrolytes made from polymer complexes formed by lithium salts with polyethers, such as poly(ethylene oxide) and poly(propylene oxide). Some of the major problems found in batteries utilizing solid polymer electrolytes can be attributed to difficulties in lowering the resistance of the solid polymer electrolytes. Because of the difficulty in decreasing their specific resistivities, it is necessary instead to make the solid polymer electrolytes very thin. Plasma polymerization is a useful method for depositing uniform, ultra-thin polymer layers on various substrates. We have reported the preparation of such thin solid polymer electrolyte films by plasma polymerization in previous communications (1,2). Here, we report on the fabrication and performance of thin-film solid-state lithium batteries (total thickness less than 20 μm) incorporating polymer

179

electrolyte films prepared by plasma polymerization. In particular, the effects of the interfaces between the solid polymer electrolyte and the electrodes, especially the product of the reaction between plasma polymer and lithium anode, on the discharge performance of solid-state batteries, are discussed in detail.

2. EXPERIMENTAL

The apparatus used to carry out the plasma polymerization has been described in detail in a previous communication (3). The apparatus consisted of a glass reactor equipped with capacitively-coupled inner electrodes to which an alternating voltage was applied at a frequency of 13.56 MHz, a monomer inlet, and vacuum system consisting of a mechanical booster pump, a rotary pump, and a cold trap. The pressure in the reactor was monitored using a Pirani gauge. Ar gas (10 cm^3 (STP)/min) and tris(2-methoxyethoxy)vinylsilane [TMVS] vapor (1 cm^3 (STP)/min) were introduced into the reactor. The pressure was reproducibly maintained at 0.3 Torr by controlling a needle valve. After setting the reactor conditions, the RF power of 5 W was turned on, and the plasma polymerization was carried out. The thin plasma polymer film was deposited on the substrate, which was placed between two electrodes. The deposition rate was about 6 um h^{-1}. The FT-IR spectra of the plasma-polymerized TMVS and poly(TMVS) both were similar to that of bulk TMVS, except for the complete absence of the characteristic absorption peak of the olefin group at 1600 cm^{-1}. This indicates the plasma-polymerized TMVS is very similar to poly(TMVS), as we also reported in an earlier communication (2).

Samples for FT-IR measurements were prepared by first depositing a layer of plasma polymer formed from TMVS of about 50-70 nm thickness on a KBr plate. A thin layer of lithium metal was deposited on the plasma polymer layer, and the sample then was maintained at 80 $°C$ for 8 h under Ar atmosphere, in order to allow the lithium to react with the plasma polymer.

A schematic view of an apparatus for FT-IR measurement of the product of the reaction between plasma polymer and lithium is shown in Fig. 1.

Lithium alkoxides and lithium alkylsilanolate as reference compounds for FT-IR are prepared as followes: 1-butyllithium was reacted with excess methanol, ethanol, butanol, and trimethylsilanol in hexane at room temperature to yield lithium methoxide, lithium ethoxide, lithium butoxide, and lithium trimethylsilanolate, respectively. The excess alcohol or silanolate and hexane used as the solvent was evaporated at room temperature under 10^{-3} Torr. The resulting lithium alkoxides and lithium alkylsilanolate were white powders.

3. RESULTS AND DISCUSSION

Many workers have reported the formation of a film on the lithium surface by reaction of the lithium and the electrolyte between the lithium and the liquid electrolyte (4-8). The interfacial layer, which is slightly conductive for lithium ion, behaves as a passivating film during the discharge of lithium batteries. The surface films formed on lithium in various organic solvents have been identified by Aurbach et. al.(9-12). For example, lithium alkoxides are formed on lithium surfaces by reaction of ether solvents, e.g., diethoxyethane and tetrahydrofuran, with lithium (10).

FT-IR transmission spectra of lithium alkoxides reference compounds, obtained using the KBr tablet method, are shown in Fig. 2. The spectra of lithium methoxide (Fig. 2a), lithium ethoxide (Fig. 2b), and lithium butoxide (Fig. 2c) are in good agreement with spectra reported in the literature (9,10). In the KBr tablet method the lithium alkoxides were in contact with KBr, and may have reacted with KBr, resulting in the formation of potassium alkoxides and LiBr. Absorption by these latter compounds interferes with and obscures absorption by the alkoxides (Li-O stretching) in the low wave number region. In order to avoid this interference the nujol mull method was also utilized. FT-IR transmission spectra of same reference compounds as in Fig. 2, but obtained by the nujol mull method, are shown in Fig. 3. The absorption peaks in Fig. 3 at 2950, 2925, 2855, 1460, 1380, and 725 cm^{-1} can be attributed to the nujol. The IR spectra of Fig. 3 are otherwise very similar to those of Fig. 2, except for the LiO stretching vibrations which occur at about 600-400 cm^{-1}. Peak assignments for FT-IR spectra of lithium alkoxides were performed using the spectra obtained by the nujol mull method, except for peaks at 3500-2500, 1500-1300, 750-700 cm^{-1} because of overlap with absorption of nujol. FT-IR transmission spectra of lithium trimethyl-silanolate obtained by the nujol mull method and the KBr tablet method are shown in Fig. 4a and b, respectively. Table. 2 shows the main peak assignments for FT-IR spectra of lithium methoxide, lithium ethoxide, lithium butoxide and lithium trimethylsilanolate.

The FT-IR transmission spectrum of the product of the reaction between the thin film of plasma polymer formed from tris(2-methoxyethoxy)vinylsilane and lithium is shown in Fig. 5. This spectrum of the thin film was obtained by the novel method described in the Experimental section above. Although the peaks of the spectrum are broad because of the high molecular weight, the spectrum is similar to those of lithium alkoxide and lithium alkylsilanolate. Table 1 shows the main peak assignments for the FT-IR spectrum of the product of the reaction between the plasma polymer and lithium with reference to the assignments for the reference compounds (Table 2). These result indicate that the product is a mixture of lithium alkoxide and lithium alkylsilanolate. Thus, the formation of a film of lithium alkoxides and lithium alkylsilanolate likely increased the resistance at the anode/electrolyte interface in the thin lithium batteries of this study. Possible paths for the reactions between plasma polymer formed from

181

tris(2-methoxyethoxy)vinylsilane and lithium are shown in Fig. 6.

REFERENCES

1) Z. Ogumi, Y. Uchimoto, Z. Takehara, and F.R. Foulkes,
   J. Electrochem. Soc., 137, 29(1990).
2) Y. Uchimoto, Z. Ogumi, Z. Takehara, and F.R. Foulkes,
   ibid, 137, 35(1990).
3) Z. Ogumi, Y. Uchimoto, and Z. Takehara, ibid, 136, 625(1989).
4) Y. Geronov and R.H. Muller, ibid, 132, 285(1985).
5) A.N. Dey, Thin Solid Films, 43, 131(1977).
6) E. Peled, J. Electrochem. Soc., 126, 2047(1979).
7) I. Yoshimatsu, T. Hirai, and J. Yamaki, ibid, 135, 2422(1988).
8) C.D. Desjardins, G.K. MacLean, and H. Sharifian, ibid, 136, 345(1989).
9) D. Aurbach, M.L. Daroux, P.W. Faguy, and E. Yeager, ibid, 134,
   1611(1987).
10) D. Aurbach, M.L. Daroux, P.W. Faguy, and E. Yeager, ibid, 135,
    1863(1988).
11) D. Aurbach, ibid, 136, 906(1989).
12) D. Aurbach, ibid, 136, 1606(1989).

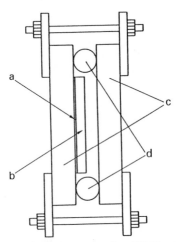

Fig. 1   A schematic view of an apparatus for FT-IR measurement of the
reaction product between the thin ultra-film plasma polymer
formed from tris(2-methoxyethoxy)vinylsilane and lithium.
a: unreacted plasma polymer layer, b: product of the reaction
between the plasma polymer and lithium, c: KBr plate,
d: O-ring.

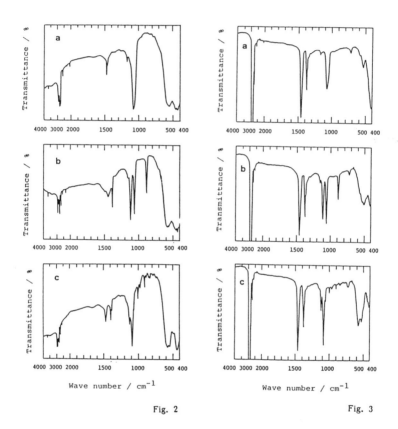

Fig. 2   FT-IR transmission spectra of (a) lithium methoxide, (b) lithium
ethoxide, (c) lithium butoxide by KBr tablet method.

Fig. 3   FT-IR transmission spectra of (a) lithium methoxide, (b) lithium
ethoxide, (c) lithium butoxide by nujol mull method.

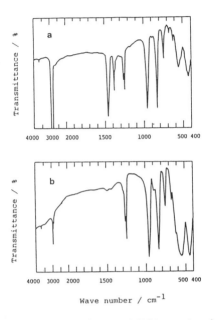

Fig. 4　FT-IR transmission spectra of lithium trimethylsilanolate by (a) nujol mull method and (b) KBr tablet method.

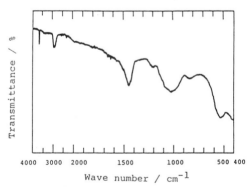

Fig. 5　FT-IR transmission spectrum of the product of the reaction between plasma polymer formed from tris(2-methoxyethoxy)-vinylsilane and lithium.

184

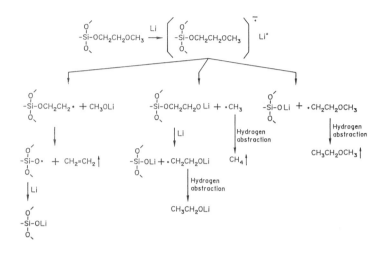

Fig. 6   Possible reaction scheme for reaction between plasma polymer formed from tris(2-methoxyethoxy)vinylsilane and lithium.

Table 1.   Main peak assignments for FT-IR spectrum of the product of the reaction between plasma polymer formed from tris(2-methoxyethoxy)vinylsilane and lithium.

Band description	Wave number / cm^{-1} product
CH stretch	2900
CH bend	1400-1500b
CO stretch ⎤	
SiO stretch ⎦	900-1100b
CH$_2$ stretch	750-900wb
LiO stretch	540b
	400b

w: weak, b: broad

185

Table 2. Main peak assignments for FT-IR spectra of lithium methoxide, lithium ethoxide, lithium butoxide and lithium trimethylsilanolate.

Wave number / cm^{-1}

Band description	lithium methoxide	lithium ethoxide	lithium butoxide	lithium trimethyl-silanolate
CH stretch	2930 2850 2800 2600w	2960 2810 2710 2610	2960 2920 2870 2810 2720	2955 2900
CH bend	1470m 1400-1500wb	1450m 1390m 1400-1600wb	1460 1360 1400-1600wb	1500m 1440m 1300-1500wb
CH_3 bend				1260s 1240s 835s
CO stretch	1160w 1075s	1120s 1060s	1120 1080s	
SiO stretch				955s
CH_2 stretch		885s	995w 970w 900w 835w	745m
LiO stretch	535 415s	535 505 420	575s 530s 420	560 430

s: strong, m: medium, w: weak, b: broad

186

CONDUCTING POLYMERS FOR CAPACITOR APPLICATION:
IN-SITU ELECTROCHEMICAL QUARTZ CRYSTAL MICROBALANCE ANALYSIS ON IONIC MOTION ACROSS CONDUCTING POLYMER CATHODE

Katsuhiko NAOI and Noboru OYAMA

Department of Applied Chemistry, Faculty of Technology
Tokyo University of Agriculture & Technology
Koganei, Tokyo 184, JAPAN

The combination of mass change and cyclic voltammetric measurements has been made for electropolymerized conducting polymer cathodes(polyaniline, polypyrrole). For the application of conducting polymer in high-power capacitor devices, it is crucial to know the mobile species(anions, cations, or protons) across electroactive thin films. The measurements were done by using in-situ electrochemical quartz crystal microbalance(EQCM). Polypyrrole films which exchanges anions showed generally slower redox, whereas the cation-exchanging polypyrrole films has relatively faster redox behavior.

INTRODUCTION

The QCM method enables us to characterize the mass change in a direct manner, namely under in-situ conditions(1) The method have several advantages that have been demonstrated for various cases including the followings: (i) metal electrodeposition and dissolution, (ii) redox cycling of conducting polymers (polypyrrole, polyaniline) and redox polymers (polyvinylferrocene).

In the present paper, the method was applied to monitor the mass change due to electrochemical polymerization and the redox reactions of various kinds of polypyrrole and polyaniline films which had been prepared with various kinds of dopants. As shown in Fig.1, the films are one of the future materials for high-power capacitor systems compared to the conventional materials(2). It is important to know the stoichiometry of ion exchange in order to estimate the minimum requirement of electrolyte to obtain the highest specific energy of the systems. The QCM mass sensors was employed to detect the insertion/ extraction of ions, either anions or cations or both, that are induced by redox cycling of the polymer films.

EXPERIMENTAL

Gold (ca.2000Å thick)-coated quartz crystals(5 or 6 MHz AT-cut) were employed as the substrate for the working electrode and subject to the mass change measurement. Onto quartz crystal, conductive polymer films were directly electropolymerized(Scheme 1). The resonance frequency shift, which has a linear relationship with the mass change was measured with a frequency counter, and the data was stored in a personal computer. Experiments were carried out under nitrogen atmosphere.

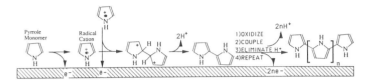

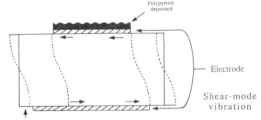

Scheme 1   Electropolymerization of pyrrole on to AT-cut quartz crystal.

The resonant frequency shift basically obeys the Sauerbrey's equation(4):

$$\Delta f = - \frac{2f_0^2}{\sqrt{r\Omega\mu\Omega}} \frac{\Delta m}{A}$$

However, the absolute mass sensitivity was determined by calibrating it with the reaction of $Ag(s) \leftrightarrow Ag^+ + e^-$.

RESULTS AND DISCUSSION

MIXED ANION & CATION EXCHANGE AT POLYPYRROLE FILMS

The authors have confirmed by QCM measurement that the polypyrrole(PPy) film prepared with large polymeric anions like PVS-, incorporated cations in the reduction process instead of releasing anions because the large anions could not be untangled from the polypyrrole matrix(2). By varying the size of anions, alkylsolfonate was used for the preparation of polypyrrole films. The mass changes due to the film formation and the subsequent reduction process were monitored by EQCM. Typical results for anion-exchange and cation-exchange behavior were shown for the electropolymerized films of PPy/C3H7SO3-(abbreviated as C3) and PPy/C5H11SO3-(abbreviated as C5) films, respectively(see Fig.2). Also, the QCM response corresponding to the pulse electropoly-merization and the subsequent reduction of the polymer films are shown in Fig.3. Between C3 and C5, the ion-exchange process of polypyrrole was converted from the anion-exchanging to cation-exchanging behavior.

DOPING MECHANISM OF POLYANILINE FILMS(4,5)

A) In Acidic Aqueous Solution:: The QCM frequency response against potential is consistent with that reported by Orata and Buttry where the polyaniline film was soaked in 1.0M H_2SO_4 aqueous solution (pH 1.75). At the first oxidation peak, the polyaniline film is oxidized to form cationic sites or cation radicals and the positive charge requires the insertion of anions for charge neutrality. Thus, the mass of the polyaniline film continues to increase during the first oxidation peak. At the second oxidation peak, on the other hand, a significant increase in frequency is observed, which indicates a mass loss on oxidation at potentials more positive than 0.55 V. The frequency change can be caused by deprotonation from the nitrogen atoms to be the imine form.

B) In Non-Aqueous Media:: In contrast, a polyaniline film soaked in non-aqueous media shows somewhat different QCM responses, which may indicate that the different doping mechanism occurs. The voltammogram shows two pairs of redox peaks as being similar to that observed in acidic aqueous solution. However, the QCM response is quite different especially at the second redox process. The frequency keeps decreasing even at the second redox peak, indicating that the mass of the polyaniline film continues to increase even at the second redox peak. Thus, the second oxidation process as well as the first must be accompanied by anion insertion (anion doping) as shown in Fig.4. Also, possible reaction schemes for the polyaniline redox both in aqueous and non-aqueous solutions were proposed.

ACKNOWLEDGEMENT

The authors wish to acknowledge the financial support from TEPCO Research Foundation and an additional support from the Japanese Ministry of Education.

REFERENCES

1. W. H. Smyrl and K. Naoi, *AIChE Symp. Series* 278, **86**, 71 (1990).
2. K. Naoi, M. M. Lien, W. H. Smyrl and B. B. Owens, *Appl. Phys. Comm.*, **9**, 147 (1989).
3. K. Naoi, M. M. Lien and W. H. Smyrl, *J. Electrochem. Soc.*, **138**, 440 (1991).
4. G. Sauerbrey, *Z. Phys.*, **155**, 206 (1959).
5. H. Daifuku, T. Kawagoe, N. Yamamoto, T. Ohsaka and N. Oyama, *J. Electroanal. Chem.*, **274**, 313 (1989).
6. H. Daifuku, T. Kawagoe, T. Matsunaga, N. Yamamoto, T. Ohsaka and N. Oyama, *Synthetic Metals*, **41**, 2897 (1991).

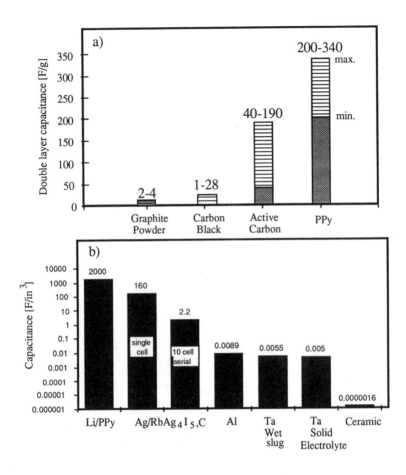

Fig.1 Comparison of the capacitance values of polypyrrole and Li/Polypyrrole cell with carbon materials or other conventional double-layer capacitor(3).

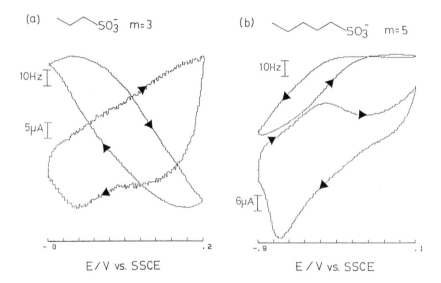

Fig.2   I-E and F-E curves for polypyrrole films(50mC/cm^2) doped with (a) $C_3H_7SO_3^-$ and (b) $C_5H_{11}SO_3^-$ anions. Measurement was done in aqueous solutions containing respective Na salts of the anions.

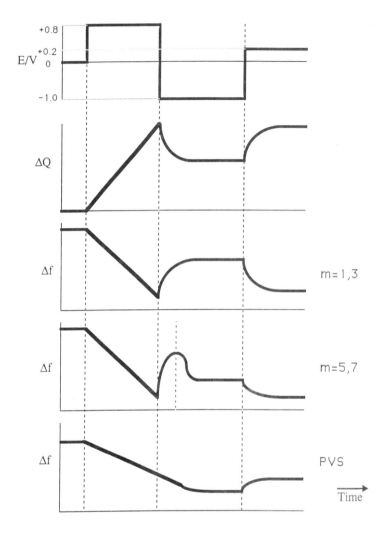

Fig.3   Typical pulse electropolymerization of pyrrole with various dopants
and their respective QCM resposes.

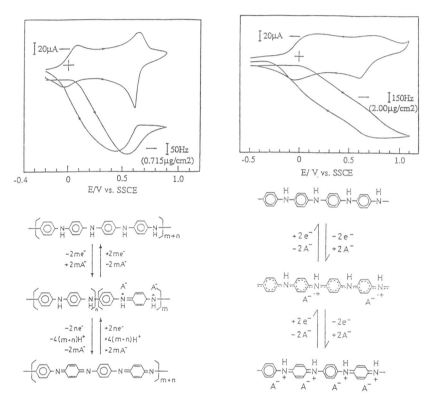

Fig.4 Cyclic voltammograms and QCM responses for a polyaniline film at Pt substrate in [LEFT] (0.5M NaClO4+HClO4)aq. (pH=1.0) and [RIGHT]1.0 M LiClO4/PC solutions. Scan rate : 1 mV/s. Charges passed during electropolymerization: 5 C/cm^2

AUTHOR INDEX

SUBJECT INDEX

RETURN   CHEMISTRY LIBRARY
Hildebrand Hall   642-3753

NOV 18 '92